MODERN GEOMETRIES

CONTEMPORARY UNDERGRADUATE MATHEMATICS SERIES
Robert J. Wisner, Editor

MATHEMATICS FOR THE LIBERAL ARTS STUDENT, SECOND EDITION, Fred
Richman, Carol Walker, and Robert J. Wisner

INTERMEDIATE ALGEBRA, Edward D. Gaughan

ALGEBRA: A PRECALCULUS COURSE, James E. Hall

TRIGONOMETRY: CIRCULAR FUNCTIONS AND THEIR APPLICATIONS,
James E. Hall

MODERN MATHEMATICS: AN ELEMENTARY APPROACH, THIRD
EDITION, Ruric E. Wheeler

A PROGRAMMED STUDY OF NUMBER SYSTEMS, Ruric E. Wheeler and
Ed R. Wheeler

FUNDAMENTAL COLLEGE MATHEMATICS: NUMBER SYSTEMS AND
INTUITIVE GEOMETRY, Ruric E. Wheeler

MODERN MATHEMATICS FOR BUSINESS STUDENTS, Ruric E. Wheeler
and W. D. Peeples

ANALYTIC GEOMETRY, James E. Hall

INTRODUCTORY GEOMETRY: AN INFORMAL APPROACH, SECOND
EDITION, James R. Smart

MODERN GEOMETRIES, James R. Smart

AN INTUITIVE APPROACH TO ELEMENTARY GEOMETRY, Beauregard
Stubblefield

GEOMETRY FOR TEACHERS, Paul B. Johnson and Carol H. Kipps

LINEAR ALGEBRA, James E. Scroggs

ESSENTIALS OF ABSTRACT ALGEBRA, Charles M. Bundrick and John J.
Leeson

AN INTRODUCTION TO ABSTRACT ALGEBRA, A. Richard Mitchell and
Roger W. Mitchell

INTRODUCTION TO ANALYSIS, Edward D. Gaughan

DIFFERENTIAL EQUATIONS AND RELATED TOPICS FOR SCIENCE AND
ENGINEERING, Robert W. Hunt

A PRIMER OF COMPLEX VARIABLES WITH AN INTRODUCTION TO
ADVANCED TECHNIQUES, Hugh J. Hamilton

CALCULUS OF SEVERAL VARIABLES, E. K. McLachlan

PROBABILITY, Donald R. Barr and Peter W. Zehna

THEORY AND EXAMPLES OF POINT-SET TOPOLOGY, John Greever

AN INTRODUCTION TO ALGEBRAIC TOPOLOGY, John W. Keesee

EXPLORATIONS IN NUMBER THEORY, Jeanne Agnew

NUMBER THEORY: AN INTRODUCTION TO ALGEBRA, Fred Richman

MODERN GEOMETRIES

JAMES R. SMART
California State University, San Jose

BROOKS/COLE PUBLISHING COMPANY
Monterey, California
A Division of Wadsworth Publishing Company, Inc.

This book was edited by Phyllis London and designed by Linda Marcetti. The technical art was drawn by John Foster. The book was printed and bound by Kingsport Press, Kingsport, Tennessee.

ISBN: 0-8185-0051-4
L.C. Catalog Card No: 72-79015
Printed in the United States of America

1 2 3 4 5 6 7 8 9 10—77 76 75 74 73

PREFACE

In recent years, the traditional course in college geometry often has been dropped, only to be replaced by other courses no more satisfactory. In some cases, the mistake has been made of throwing out all of Euclidean geometry, both traditional and modern, whether it is of continuing significance or not. The even worse mistake is sometimes made of assuming that students understand geometry simply because they have taken some nongeometric course such as linear algebra.

I believe that the traditional course in college geometry should be replaced by an extensive course in modern geometry that includes significant topics from modern Euclidean geometry. This text provides adequate material for such a two-semester or three-quarter course at the upper-division college level. The scope is broad but not all-inclusive.

Material is presented from foundations of geometry, finite geometry, the modern Euclidean geometry of convexity, modern college geometry (including constructions), and transformations. There are separate chapters on projective geometry, non-Euclidean geometry, inversive geometry, and geometric topology. Within the various chapters are such new and intriguing topics as caroms and Morley's theorem, as well as fascinating geometry stemming from consideration of the golden ratio. Throughout, the text emphasizes practical and up-to-date applications of modern geometry. The student should be aware that many of these topics are discussed in current professional journals and that contemporary research mathematicians are seriously involved in the extension of geometric ideas.

Most of the geometry included in this text is synthetic rather than analytic, but many important ideas from analytic geometry are used as needed. An introduction to the analytic approach is provided for transformations and for projective and inversive geometry.

The text is not a sweeping survey. Instead, each major topic is developed so that the student can become quite knowledgeable about the procedures, concepts, and proofs involved in the various geometries. The chapters actually present concise courses on each kind of geometry, not just summaries of what each type is about.

The text is written for students who range widely in their mathematical abilities. Much of the material is appropriate for those who have average or weak backgrounds in geometry. On the other hand, students with strong backgrounds will also find much to interest them. The text is planned for both majors and minors in mathematics. It is appropriate for students interested in mathematics from the liberal arts standpoint and for those planning to be teachers of mathematics.

Many of the first exercises in each set can be used orally as the basis for classroom discussion; in that way the instructor can make certain that fundamental concepts are understood. Later exercises allow extensive practice in providing independent proofs.

The text encourages independent investigation so that the student will have experience in using geometric intuition, making conjectures, and then proving formal theorems. Advanced theorems are sometimes inserted without proof if they contribute to the development of a topic.

Because good teachers often prefer to bring in necessary review topics when they are needed rather than in a single introductory chapter, the text includes review at appropriate intervals. Students should find a high school geometry text useful when a reference is needed.

The various chapters of the text are largely independent. This and the arrangement of topics within each chapter allow for great flexibility in their use, according to the needs of the class and the desires of the instructor. The teacher who wishes only minimal coverage may use just the first sections of a chapter. The later sections may be used for a more complete and rigorous course if time allows.

While the entire text provides adequate material for a two-semester or three-quarter course in modern geometry, sections of it are appropriate for use in shorter courses. Some suggestions for such use are given below.

1. Two-quarter course in survey of modern geometry:
   All of text except for Chapters 4 and 5; also possibly omit Sections 3.6 and 7.8–7.10 in Chapters 3 and 7.
2. One-semester course in Euclidean geometry:
   All sections through Section 5.3.
3. One-semester course in survey of modern geometry:
   Finite geometries, Chapters 2, 3, 7, 8.
4. One-semester course in modern geometry, not including Euclidean geometry:
   Finite geometries, Chapters 6, 7, 8, 9.
5. One-quarter course in Euclidean geometry:
   All sections through 5.3, possibly omitting Sections 1.5, 1.6, 2.6, 3.6, and 4.6.
6. One-quarter course in survey of modern geometry:
   Finite geometries, Chapters 2, 3, 7, 8, possibly omitting Sections 2.6, 3.6, and 7.8–7.10.
7. One-quarter course in modern geometry, not including Euclidean geometry:
   Finite geometries, Chapters 7, 8, 9.

I would like to express my appreciation to James Moser, of the University of Wisconsin, Bruce Partner, of Ball State University, John Peterson, of Brigham Young University, Demitrios Prekeges, of Eastern Washington State College, Curtis Shaw, of the University of Southwestern Louisiana, and Marvin Winzenread, of California State University, Hayward, for reviewing the manuscript, and to the many serious college students who made worthwhile suggestions for improvements while the material was being used in preliminary form.

CONTENTS

CHAPTER 1

SETS OF AXIOMS AND FINITE GEOMETRIES

1.1 INTRODUCTION

The word "geometry" can be interpreted literally as "earth measure," but this literal meaning seems far removed from the various modern geometries to be explored in this text. The idea of earth measure arises from the ancient, pre-Greek development of geometry. These practical Egyptian and Babylonian applications of geometry involved measurement to a great extent, and they were not complicated by formal proofs. For example, the properties of the right triangle were known to the extent that a rope with knots (Figure 1.1), held tightly by three men so as to form a right triangle, could be used in surveying.

FIGURE 1.1

During the Greek period, the science of earth measure became more refined. About 230 B.C., Eratosthenes made a remarkably precise measurement of the size of the earth. According to the familiar story, Eratosthenes knew that at the summer solstice the sun shone directly into a well at Syene at noon. He found that at the same time, in Alexandria, approximately 489 miles due north of Syene, the rays of the sun were inclined about 7.2° from the vertical (Figure 1.2). With these measurements, Eratosthenes was able to find the diameter of the earth.

Interestingly enough, the earth-measurement aspect of geometry has been of recent interest because of the importance of measurements of great precision made by satellites, by instruments placed on the moon, and by the U.S. Coast and Geodetic Survey in producing nautical and aeronautical charts.

FIGURE 1.2

The ancient Greeks of the period 500 B.C. to 100 A.D. receive much of the credit for the development of demonstrative geometry of the sort studied at the high school level. They recognized the beauty of geometry as a discipline with a structure and understood that the proof of a theorem could be even more exciting than the discovery of a practical application. The geometry of Chapters 2 through 5 is essentially based on these early Greek contributions, though most of it was developed much later.

Greek geometry, called *Euclidean geometry* because of the monumental work of Euclid (300 B.C.), includes *undefined terms, defined terms, axioms* or *postulates,* and *theorems.* Almost every geometry studied in this text has the same sort of structure. In Euclidean geometry, undefined terms, which are arbitrary and could easily be replaced by other terms, normally include points, lines, and planes; it would be possible to develop Euclidean geometry using such concepts as distance and angle as undefined. Definitions of new words involve use of the undefined terms. The fact that definitions usually contain undefined terms is standard in mathematics.

Today, the words *axiom* and *postulate* are used interchangeably. In the development of geometry, however, the word *postulate* was used for an assumption confined to one particular subject (such as geometry), while *axiom* denoted a "universal truth," a more general assumption that applied to all of mathematics. The axioms and postulates of Euclid are stated in Section 1.2. The *truth* of axioms or postulates is not at issue. These statements are beginning assumptions from which logical consequences follow. They are analogous to the rules for a game. Since the mathematical system to be developed depends on the axioms, changing the axioms can greatly change the system, just as changing the rules for a game would change the game.

Theorems are statements to be proved by using the axioms, definitions, and previous theorems as reasons for the logical steps in the proof. The theorems of geometry are valid conclusions based on the axioms. A simple theorem typically is stated in the form of an if-then statement such as "If the sum of the measures of the opposite angles of a quadrilateral is 180 (in degrees), then the quadrilateral can be inscribed in a circle." In logic, this theorem is an *implication.* The hypothesis (the "if" part) is often identified as the statement p and the conclusion (the "then" part) may be identified as the statement q. The implication is "if p, then q," or $p \rightarrow q$ (p implies q). The proof of a theorem means a proof of the entire implication, not just proof of the conclusion. A deductive proof begins with the hypothesis and leads through a series of logical steps to the conclusion. Up to this point, the proof has shown that the conclusion q is always true if the hypothesis p is true. For this reason, it is important to realize that the deduction theorem from logic is involved whether or not it is stated explicitly in the steps of the proof.

DEDUCTION THEOREM. If, under a given true premise p, the conclusion q is always true, then the entire statement p implies q is true.

Since many examples of theorems and proofs will be given in later sections, the concepts will become more meaningful with experience.

The discussion above has provided a description of some of the basic premises and methods of synthetic geometry. Really significant advances over the synthetic geometry of the Greeks were made only with the invention of analytic geometry (about 1637) and its subsequent use as a tool in modern analysis. While analytic geometry is not the dominant theme of this text, coordinates of points are used as an alternative to the synthetic approach when convenient.

As the title *Modern Geometries* implies, the major emphasis is on newer geometries that have been developed since 1800. The emergence of modern algebra, with its theory of groups, and the introduction of axiomatics into algebra paved the way for Felix Klein's classification of geometries in 1872. The basic concept of transformations needed to understand this classification is discussed in Chapter 2. The latter part of the nineteenth century witnessed a revival of interest in the classical geometry of the circle and the triangle, with the result that the Greek geometry was extended by many significant additions (Chapters 4 and 5). Projective geometry (Chapter 7) was invented about 1822; material on non-Euclidean geometry (Chapter 9) was in print by about 1830. Inversive geometry (Chapter 6) was developed about the same time.

During the twentieth century, studies in the axiomatic foundations of geometry and the finite geometries (Chapter 1), the geometry of convexity (Chapter 3), and geometric topology (Chapter 8) have all been added to the great body of geometry that is relatively independent of analysis. Although they are outside the scope of this text, differential geometry and vector analysis, to name but two areas of investigation, could also be covered in the study of modern geometry.

Even this brief sketch of some of the major steps in the history of geometry should have convinced you that a discussion of modern geometries must deal with many different kinds of geometry. It is the acknowledgement of the diversity of mathematical systems deserving the title of geometry that distinguishes a book on modern geometries from a traditional college geometry text of a quarter of a century ago, which concentrated only on a direct extension of the Euclidean geometry of the high school.

EXERCISES 1.1

(Answers to selected exercises are given at the back of the text.)

1. a. Verify that the Pythagorean theorem, $a^2 + b^2 = c^2$, holds for the sides of the triangle in Fig. 1.1.
    b. How many knots in the rope would be necessary to produce a triangle with shorter sides measuring 5 and 12 units?

2. Use the measurements of Eratosthenes to find the approximate diameter of the earth.

For Exercises 3–8, answer true or false, then explain what is wrong with each false statement.

3. High school geometry owes more to the ancient Egyptians than to the ancient Greeks.

4. Euclid used the word *postulate* for an assumption confined to one particular subject.

5. The deduction theorem is used because a conclusion, not an implication, is to be proved.

6. Analytic geometry was invented before the development of finite geometries.

7. The latter part of the nineteenth century witnessed a revival of interest in the classical geometry of the circle and the triangle.

8. Traditional college geometry of a quarter of a century ago included the study of more different geometries than are included today.

1.2 SETS OF AXIOMS FOR EUCLIDEAN GEOMETRY

The importance of a wise choice of axioms was implied in the previous section, but the selection of axioms is no simple matter. A major problem in geometry for some 2000 years was how to provide an adequate set of axioms for ordinary Euclidean geometry. Several such sets of axioms are discussed in this section.

Euclid, about 300 B.C., gave a famous set of five axioms and five postulates as follows:

Axioms (or common notions)

1. Things that are equal to the same thing are also equal to one another.
2. If equals are added to equals, the wholes are equal.
3. If equals are subtracted from equals, the remainders are equal.
4. Things that coincide with one another are equal to one another.
5. The whole is greater than the part.

Postulates

1. A straight line can be drawn from any point to any point.
2. A finite straight line can be produced continuously in a straight line.
3. A circle may be described with any point as center and any distance as radius.
4. All right angles are equal to one another.
5. If a transversal falls on two lines in such a way that the interior angles on one side of the transversal are less than two right angles, then the lines meet on that side on which the angles are less than two right angles.

Postulate 2 means in essence that a segment can be extended indefinitely to form a line. The *fifth postulate* of Euclid (the parallel postulate) will be considered further in Chapter 9 in the discussion of non-Euclidean geometry.

Within the past hundred years in particular, mathematicians studying the foundations of mathematics have pointed out various flaws in the assumptions of Euclid. Euclid actually used other, tacit, assumptions (assumptions not stated explicitly). Logical problems pointed out have included:

    a. The need for a definite statement about the continuity of lines and circles.

    b. The need for a statement about the infinite extent of a straight line.

c. The need to state the fact that if a straight line enters a triangle at a vertex, it must intersect the opposite side.

d. The need for statements about the order of points on a line.

e. The need for a statement about the concept of betweenness.

f. The need for a statement guaranteeing the uniqueness of a line joining two distinct points.

g. The need for a more logical approach, such as that of transformations (Chapter 2), which does not depend on the concept of *superposition.* Euclid assumed that a triangle can be picked up and put down in another place with all the properties remaining invariant, yet no statement to this effect was made.

h. The need for a list of undefined terms.

Many modern sets of axioms for Euclidean geometry have been introduced to remedy the defects in Euclid. In a course on the foundations of geometry, or on the foundations of mathematics, these are studied extensively. In general, these newer sets of axioms were more comprehensive than that of Euclid, but for this reason they appear more complex. It has been difficult to develop the beginning high school geometry course without resorting to the device of introducing still more axioms to avoid the proof of very difficult theorems at the beginning of the course.

One of the first modern sets of axioms for Euclidean geometry was devised by Moritz Pasch in 1882. He is given credit for what is called *Pasch's axiom*: A line entering a triangle at a vertex intersects the opposite side. Guiseppi Peano provided another new approach in 1889. Probably the most famous set of axioms for Euclidean geometry was given by David Hilbert and was published in English in 1902. These axioms are given in Appendix 1. Hilbert used six undefined terms: *point, line, plane, between, congruent,* and *on.* He gave his axioms in five groups, and the grouping helps show how the logical difficulties of Euclid were overcome. For example, note that Hilbert's first three axioms clarify Euclid's postulate about drawing a straight line from any point to any point and that his axioms of congruence take care of the logical defect of superposition.

Since the time of Hilbert there have been many other sets of modern postulates for Euclidean geometry. Some of these are by Oswald Veblen (in 1904 and 1911), Huntington (1913), Henry Forder (1927), G. D. Birkhoff (1932), and others. The axioms of Birkhoff (Appendix 2) are of special significance because they emphasize the connections between geometry and the real numbers, because they include *distance* and *angle* as undefined terms, and because they have been modified for incorporation in most recent experimental geometry courses, such as that of the School Mathematics Study Group, at the secondary school level.

One of the main reasons why Birkhoff's system is so brief is the power of the first postulate. Since this assumption has the effect of assuming all the properties of the real numbers, the order relations for points depend only on theorems and definitions rather than on additional postulates.

Because modern geometry programs depend on the coordination of analytic geometry with synthetic geometry more than was the case even at the time of Hilbert, axioms for the real numbers and the logical assumptions of algebra for relations are needed in geometry. Sets of axioms for modern geometry texts ordinarily contain specific axioms needed for each of the following purposes not readily apparent as a result of the axioms of Hilbert:

a. To show the existence of a correspondence that associates a unique number with every pair of distinct points.

b. To establish the measure of the distance between any two points in the line as the absolute value of the difference of their corresponding numbers.

c. To state the existence of a unique coordinate system for a line that assigns to two distinct points two given distinct real numbers.

d. To formulate the logical assumption necessary for the development of the theory of convexity (Chapter 3) by stating that a line in a plane partitions the points of the plane not on the line into two convex sets such that every segment that joins a point of one set to a point of the other intersects the line.

e. To include additional assumptions about congruence of triangles—assumptions that eliminate the need for lengthy proofs at an early stage in the text. The two most common additional axioms are the assumptions that (1) congruence of triangles follows from congruence of two angles and the included side and that (2) congruence of triangles follows from congruence of the three sides.

f. To postulate additional assumptions about fundamental concepts of area. These include the existence of a correspondence that associates the number 1 with a certain polygonal region and a unique positive real number with every convex polygonal region; a statement that if two triangles are congruent, the respective triangular regions have the same area; and the assumption that the measure of area of a rectangular region is the product of the measures of the lengths of its base and altitude.

An example of a modern set of axioms for secondary school geometry appears in Appendix 3.

Sets of axioms for Euclidean geometry, as well as for any mathematical system, should have two important properties.

1. The set should be *consistent*. In a consistent set of axioms, it is not possible to use them to prove a theorem that contradicts any axiom or other theorem that has already been proved. All of the sets of axioms in this text are examples of consistent systems. The concept of consistency becomes more significant after the discussion in Chapter 9 of the consistency of non-Euclidean geometry. An example of an inconsistent system could be one that included *both* of these axioms: a. Two distinct points determine exactly one line; b. Two distinct points determine exactly two lines.

2. The set should be *complete*. It should be impossible to add a consistent, independent (see definition below) axiom to the set without introducing new undefined terms.

In the study of foundations of mathematics, it is often important that the set of axioms have at least one additional property so that none of the axioms can be proved from the remaining set of axioms. Sets of axioms having this property are called *independent sets* of axioms. All of the sets of axioms in the remaining sections of this chapter are examples of independent sets of axioms. The requirement of independence is not always desirable at a more elementary level. Therefore, secondary geometry texts ordinarily have additional axioms that could be proved from the others. These axioms are included because they are convenient to state and use early in the course or because the proofs are too difficult for that level. The assumptions of Euclidean geometry, along with some theorems proved in an introductory course, are the basic assumptions used in Chapters 2 through 5 of this text. However, you won't be expected to remember each theorem.

In order to begin the study of the application of specific sets of axioms to build geometries, it is helpful to introduce a type of geometry that displays as simple a structure as possible. In the rest of this chapter, the geometries chosen for this purpose are finite geometries, explained in the next sections.

EXERCISES 1.2

1. Draw a figure to explain the wording of Euclid's fifth postulate.

Draw a figure to explain the following axioms of Hilbert.

2. Order Axiom 2.

3. Order Axiom 4.

4. Order Axiom 5.

5. Congruence Axiom 3.

6. Congruence Axiom 4.

7. The axiom of continuity.

For Exercises 8–10, name the axioms of Hilbert that:

8. State Pasch's axiom.

9. Guarantee the uniqueness of a line joining two distinct points.

10. Deal with betweenness for points on a line.

For Exercises 11–16, state which axioms from Appendix 3:

11. Show the existence of a correspondence that associates a unique number with every pair of distinct points.

12. Establish the measure of the distance between any two points in a line.

13. State the existence of a coordinate system for a line.

14. Introduce the concept of convexity.

15. Provide assumptions about congruence of triangles.

16. Provide assumptions about the area of plane regions.

For Exercises 17–20, state whether a set of axioms could be:

17. Complete but not independent?

18. Independent but not complete?

19. Independent but not consistent?

20. Consistent but not independent?

1.3   INTRODUCTION TO FINITE GEOMETRIES

From the standpoint of structure, Euclidean geometry is complex. The Euclidean plane has an infinite number of points and lines in it, and a rich collection of theorems continues to increase over the years. By contrast, "miniature" geometries have just a few axioms and theorems and a definite number of elements that can be named by a counting number. These geometries are *finite geometries,* and they provide excellent opportunities for study of geometries with a simple structure.

All of the geometries studied in this text have a finite number of axioms and a finite number of undefined terms. Thus those features do not make a geometry finite. Instead, a finite geometry has a finite number of elements—that is, points or lines or "things to work with." For the geometries studied in this chapter, these elements can be considered points and lines. It would seem that finite geometries are thus inherently simpler than geometries with an infinite number of points and lines, although that may not be your opinion when you first encounter them.

Historically, the first finite geometry to be considered was a three-dimensional geometry, each plane of which contained seven points and seven lines. The modernity of finite geometries is emphasized by the fact that Fano explored this first finite geometry in 1892. It was not until 1906 that finite projective geometries were studied by Veblen and Bussey. Since that time, a great many finite geometries have been or are being studied. Many sets of points and lines that were already familiar figures in Euclidean geometry were investigated from this new point of view. However, at the present time it is quite possible for a mathematics major to graduate without ever encountering finite geometries, although it is also true that finite geometries are being used increasingly as enrichment topics and extension units at the high school level. Finite geometries also find a practical application in statistics.

All of the finite geometries in this chapter have *point* and *line* as undefined terms. The connotation of line is not the same in finite geometry as in ordinary Euclidean geometry, however, since a line in finite geometry cannot have an infinite number of points on it.

The first simple finite geometry to be investigated, called a *three-point geometry* here for identification, has only four axioms:

Axioms for Three-Point Geometry

1. There exist exactly three distinct points in the geometry.
2. Two distinct points are on exactly one line.
3. Not all the points of the geometry are on the same line.
4. Two distinct lines are on at least one point.

Some immediate questions to consider intuitively before reading farther are the following:

    a. What kinds of figures or models could be drawn to represent the geometry?

    b. How many lines are in the geometry?

    c. Can any theorems be proved for the geometry?

    d. What representations are possible for the geometry, other than those with points and lines?

    e. Which properties of Euclidean geometry continue to hold in the three-point geometry, and which do not?

Not all of these questions can be answered completely for each finite geometry studied, but the questions do illustrate the nature of inquiry about a geometry based on an axiomatic system. It will be immediately helpful to give a partial answer to question a.

The finite geometry of three points can be represented by many drawings, four of which are shown in Figure 1.3. Verify the fact that all four axioms hold for each figure.

FIGURE 1.3

While the sets of points and lines in Figure 1.3 are such that all the axioms of the three-point geometry hold, there is still the possibility that the geometry might have additional lines not shown. This matter is settled by proving two theorems. First, a comparison of the wording of Axioms 2 and 4 leads to the need to determine whether two distinct lines might be on more than one point.

THEOREM 1.1. Two distinct lines are on exactly one point.

By Axiom 4, two distinct lines are on at least one point. Assume that two lines are on more than one point. If two distinct lines *l* and *m* lie on points *P* and *Q*, then Axiom 2 is contradicted, because points *P* and *Q* would be on two distinct lines.

Theorem 1.1 is proved by what is called an *indirect argument*. The theorem could be rewritten in the form of an implication. If two lines are distinct, then they are on exactly one point. The assumption was made that the conclusion was not true, and a contradiction was reached, showing that the negative assumption was not tenable. Thus, the conclusion is valid, and, by the deduction theorem, Theorem 1.1 is established. Indirect proofs, which are probably more effective in geometry than in algebra, will be used many times in this text.

The exact number of lines in the three-point geometry can now be determined.

THEOREM 1.2. The three-point geometry has exactly three lines.

From Axiom 2, each pair of points is on exactly one line. Each possible pair of points is on a distinct line, so the geometry has at least three lines. Suppose there is a fourth line. From Axiom 1, there are only the three points in the geometry. This fourth line must also be on two of the three points, but this contradicts Axiom 2 and Theorem 1.1. There can be no more than three lines in the geometry.

While point and line have been used as the undefined terms in this first finite geometry, other words could be substituted to give an equally meaningful interpretation of the structure. For example, *tree* could be substituted for *point,* and *row* for *line,* so that the postulates would read as follows:

    a. There exist exactly three distinct trees.
    b. Two distinct trees are on exactly one row.
    c. Not all trees are on the same row.
    d. Two distinct rows have at least one tree in common.

Other interpretations could be found by using pairs of words such as beads and wires, students and committees, or books and libraries.

In the finite geometry under consideration, it should be evident from an examination of the axioms and Figure 1.3 that such Euclidean concepts as length of a segment, measure of an angle, and area—in fact all concepts concerning measurement—no longer apply. In this geometry, if a triangle is defined as three distinct lines meeting by pairs in three distinct points, not all collinear, then one and only one triangle exists. The concept of parallel lines does not apply, if parallel lines are defined as two lines with no points in common, since each two lines meet in a point. The familiar ideas of congruence also have no meaning in this geometry. Even though you are far more familiar with Euclidean geometry than with finite geometry, this consideration of whether the Euclidean properties hold in a new geometry will nevertheless give added meaning to the familiar concepts. The introduction of more significant finite geometries in the next sections will provide additional opportunities for the same kind of consideration.

EXERCISES   1.3

For the three-point geometry:

1. Draw a pictorial representation different from those in Figure 1.3.
2. Rewrite the axioms, using the words *book* for *point* and *library* for *line*.
3. Rewrite the axioms, using the words *student* for *point* and *committee* for *line*.
4. Through a point not on a given line, there are how many lines parallel to the given line?
5. Exactly how many points are on each line?
6. Must lines be straight in the Euclidean sense?
7. Could three lines all contain the same point?
8. Do any squares exist?
9. Prove that a line cannot contain three distinct points.

1.4   FOUR-LINE AND FOUR-POINT GEOMETRIES

The three-point geometry of Section 1.3 had a structure so simple that it seemed of relatively little significance. The two related finite geometries in this section are more rewarding to study, yet their sets of axioms are even briefer.

The next finite geometry to be considered also has points and lines as undefined terms. The following three axioms completely characterize the geometry, called a *four-line geometry* here for identification.

Axioms for Four-Line Geometry

1. The total number of lines is four.
2. Each pair of lines has exactly one point in common.
3. Each point is on exactly two lines.

Axiom 1 is an *existence* axiom, because it guarantees that the geometry is not the empty set of points. The other axioms are *incidence* axioms, dealing with points on lines and lines on points. Before reading on, try to draw diagrams of points and lines that will satisfy all of the three axioms. Also, try to determine the total number of points.

THEOREM 1.3. The four-line geometry has exactly six points.

By Axiom 1, there are six pairs of lines. The number six is obtained as the combination of four things taken two at a time. Three notations used for combinations are

$$C_{4,2}, \quad {}_4C_2, \quad \text{or} \quad \binom{4}{2}.$$

The general formula for the combination of n things taken r at a time is

$$\frac{n!}{(n-r)!r!}.$$

By Axiom 2, each pair of lines has exactly one point in common. Suppose that two of these six points are not distinct. That would be a contradiction of Axiom 3, because each point would be on more than two lines. Also, by Axiom 3, no other point could exist in the geometry other than those six on the pairs of lines.

THEOREM 1.4. Each line of the four-line geometry has exactly three points on it.

By Axiom 2, each line of the geometry has a distinct point in common with each of the other three lines, and all three of these points are on the given line. Suppose there were a fourth point on the given line. Then by Axiom 3, it must also be on one of the other lines. But this is impossible, because the other three lines already determine exactly one point with the given line, and by Axiom 2, they can only determine one. Thus, each line of the geometry has exactly three points on it.

Figure 1.4 shows two diagrams that can be used to represent this finite geometry of four lines and six points. An examination of Figure 1.4 will lead to further inquiries about the four-line geometry. For example, the following questions are typical of those that should be answered:

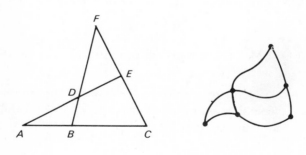

FIGURE 1.4

a. Do each two points of the geometry lie on a line?
b. How many triangles exist in the geometry (all three sides must be lines of the geometry)?
c. Does the geometry have examples of parallel lines (lines with no point in common)?

These questions are considered in the exercises at the end of the section.

The general concept of *duality* is considered in Chapter 7, but a specific example is needed here in order to explain how the next set of axioms is formulated. The *plane dual* of a statement is formed by exchanging the words *point* and *line* and making other necessary changes in the English as required. Writing the plane dual of each axiom for the previous four-line geometry results in a *four-point geometry*.

Axioms for Four-Point Geometry

1. The total number of points in this geometry is four.
2. Each pair of points has exactly one line in common.
3. Each line is on exactly two points.

Two possible representations for this geometry are shown in Figure 1.5.

FIGURE 1.5

One thing that should be noted in Figure 1.5 is that the lines meet only where points are indicated, not just where they appear to cross in the picture, because these are not ordinary Euclidean lines. If it is assumed that the four-point geometry and the four-line geometry are related in such a way that the plane dual of any valid theorem in one geometry becomes a valid theorem in the other, then it is possible to gain more information about the new geometry rather easily. Thus, the plane duals of Theorems 1.3 and 1.4 become theorems for the four-point geometry.

THEOREM 1.5. The four-point geometry has exactly six lines.

THEOREM 1.6. Each point of the four-point geometry has exactly three lines on it.

Lines exist in the four-point geometry that do not have one of the four points in common, so these lines may be considered parallel. Other properties of the four-point geometry are investigated in the exercises.

For each of the finite geometries so far, one axiom stated the exact number of points on a line or gave the total number of points or lines for the geometry. Without this limiting axiom, the sets of axioms might have resulted in geometries with an infinitude of points and lines. Indeed, most of the axioms in sets for finite geometries are also valid axioms in Euclidean geometry. For example, Axioms 2 and 3 for the finite geometry of three points hold in ordinary Euclidean geometry, and even Axiom 4 is true except when the lines are parallel. The situation is similar for other finite geometries.

EXERCISES   1.4

1. Write the plane dual of the axioms for the three-point geometry of Section 1.3.

For the four-line geometry:

2. Draw another representation for this geometry different from those shown in Figure 1.4.

3. Which axioms are also true statements in Euclidean geometry?

4. Rewrite the set of axioms for this geometry, using *student* for point and *committee* for line.

5. Do each two points of the geometry lie on a line?

6. How many triangles exist in the geometry in which all three sides are lines of the geometry?

7. How many other lines are parallel to each line?

For the four-point geometry:

8. Draw another model for this geometry different from those shown in Figure 1.5.

9. Which axioms are also true statements in Euclidean geometry?

10. Rewrite the set of axioms for this geometry, using *tree* for point and *row* for line.

11. Do each two lines of this geometry determine a point?

12. How many other lines of the geometry are parallel to each line?

13. Prove, without using the idea of duality, that the geometry includes exactly six lines.

14. Prove, without using the idea of duality, that each point of the geometry is on exactly three lines.

1.5   FINITE GEOMETRIES OF FANO
AND PAPPUS

The original finite geometry of Fano was a three-dimensional geometry, but the cross section formed by a plane passing through his configuration yields a plane finite geometry, also called *Fano's geometry,* that can be studied here. The complete set of axioms follows:

Axioms for Fano's Geometry

1.   There exists at least one line.
2.   Every line of the geometry has exactly three points on it.
3.   Not all points of the geometry are on the same line.
4.   For two distinct points, there exists exactly one line on both of them.
5.   Each two lines have at least one point in common.

For this geometry of Fano, point and line are undefined. As in previous finite geometries, the meaning of *on* in the axioms is left to the intuition. In finite geometries as well as in ordinary Euclidean geometry, various expressions can be used for the same idea. For example, all of these state the same relationship:

A point is *on* a line.

The line *contains* the point.

The line *goes through* the point.

An almost immediate consequence of the axioms is:

THEOREM 1.7.   Each two lines have exactly one point in common.

By Axiom 5, two lines have at least one point in common. The assumption that they have two distinct points in common violates Axiom 4, because then the two distinct points would have two lines containing both of them.

For the geometry of Fano, the number of points and lines is more difficult to guess. Try to guess before reading the development leading to the next theorem.

From Axioms 1 and 2, there are at least three points in the geometry, while from Axiom 3 there is at least a fourth point, as symbolized in Figure 1.6a. By Axiom 4, there must be lines joining this fourth point and each of the existing points (Figure 1.6b), and by Axioms 4 and 5, there must be lines joining points 1, 6, 7, points 3, 6, 5, and points 5, 2, 7 (Figure 1.6c). Thus the geometry of Fano contains at least seven points and seven lines.

FIGURE 1.6

The fact that there are exactly seven points in Fano's geometry can be established by an indirect argument, since the assumption of an eighth point leads to a contradiction, as will now be shown. Assume that there is an eighth point, and consider for example, as in Figure 1.7, the intersection of the line through points 1, 8 and the line 3, 7, 4. (The notation "line 3, 7, 4" means the line containing points 3, 7, and 4.) Axiom 5 requires that lines 1, 8 and 3, 7, 4 have a point of intersection. The point of intersection required by Axiom 5 cannot be point 3, 7, or 4, since that would violate Axiom 4. Thus, it must be a ninth point, but that violates Axiom 2. The assumption of an eighth point has led to a contradiction and must be rejected. The result is the following theorem:

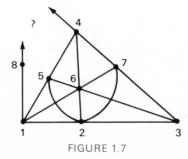

FIGURE 1.7

THEOREM 1.8. Fano's geometry consists of exactly seven points and seven lines.

Note in Figures 1.6c and 1.7 that lines in finite geometry may appear to cross without actually having a point in common. They are not Euclidean lines, and the distinct points on them need to be clearly shown in drawings to avoid confusion.

Consider a rewriting of the set of axioms for Fano's geometry with the word *point* replaced by *student* and the word *line* by *committee*. A table may be used to represent this finite system. For example, Table 1.1 shows committees in vertical columns. Substitution of numbers for the student names would show that Fano's geometry itself could be represented by the same sort of table. You should be able to set up a correspondence between each column in this table and each line in Figure 1.6c, noting that the students can be matched in alphabetical order with the points in numerical order.

TABLE 1.1

Committee 1	Committee 2	Committee 3	Committee 4
Alice	Alice	Alice	Brad
Brad	Dale	Frank	Dale
Cathy	Ellie	Greg	Frank
Committee 5	Committee 6	Committee 7	
Cathy	Brad	Cathy	
Ellie	Ellie	Dale	
Frank	Greg	Greg	

The connections between the theory of combinations and the number of points and lines in finite geometries are profitable to explore. For example, it might appear at first glance that a quick way to determine the number of lines in Fano's geometry would be to count all possible combinations of seven things taken three at a time, since each line has three points. But there are 35 combinations of seven things taken three at a time, and there are only seven lines in the geometry. A study of Figure 1.7 (or 1.6c) will make it possible to reconcile this difference. Consider points 1 and 5 and the line they determine, for example. There are five other possible points, 2, 3, 4, 6, 7, that could be matched with the two given points to form lines, and all of these possibilities are counted in the 35. But only one of these points, point 4, is actually on the line through points 1 and 5; hence, only $\frac{1}{5}$ of the actual number of possibilities result in lines. Since $\frac{35}{5} = 7$, there are a total of seven lines. Further insight can be gained by asking why the other four points, 2, 3, 6, 7, are not matched with points 1 and 5 to determine lines of the geometry. For any point such as 1, there are three lines of the geometry on it. These contain the other six points of the geometry (two on each line). None of the pairs of points on one of these lines can be matched with a point on another line without contradicting Axiom 4.

The concept of parallelism is not in evidence in Fano's geometry, since each pair of lines has a point in common. On the other hand, it would be quite possible to give a new interpretation of parallel for this finite geometry so that the concept could be considered in a way quite different from that used in Euclidean geometry. For example, suppose (using Figure 1.6c) that any two lines intersecting on line 4, 7, 3 are called parallel in Fano's geometry. Then lines 5, 2, 7 and 1, 6, 7, for example, are parallel in this interpretation, since they have point 7 on line 4, 7, 3 in common. This example helps to emphasize the important role of arbitrary definitions in determining the structure of a geometry.

Some finite geometries, though quite modern, relate to sets of

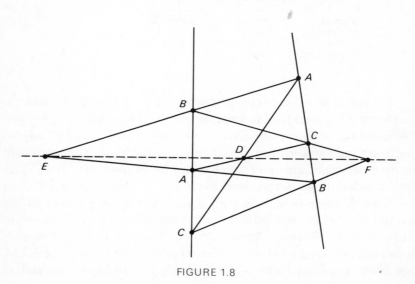

FIGURE 1.8

points and lines that have long been familiar to mathematicians. This familiarity is especially evident in the finite geometry arising from a Euclidean-geometry theorem called the Theorem of Pappus. Figure 1.8 illustrates the theorem, which was discovered and proved by Pappus of Alexandria about 340 A.D. The theorem is stated here without proof. The lines in this theorem are considered to be the same as lines in ordinary Euclidean geometry.

THEOREM 1.9. *Theorem of Pappus*: If *A, B*, and *C* are three distinct points on one line and if *A', B'*, and *C'* are three different distinct points on a second line, then the intersections of $\overleftrightarrow{AC'}$ and $\overleftrightarrow{CA'}$, $\overleftrightarrow{AB'}$ and $\overleftrightarrow{BA'}$, and $\overleftrightarrow{BC'}$ and $\overleftrightarrow{CB'}$ are collinear.

The notation $\overleftrightarrow{AC'}$ means "the line containing points *A* and *C'*." (A summary of the technical notation used here and throughout this text may be found in Appendix 4.) Points are called collinear if they lie on the same line. There are exceptions to the theorem if some of the lines are parallel (in the ordinary Euclidean sense), but it is assumed here that the lines intersect as required in real points. The theorem of Pappus is seen to involve nine distinct points, lying by threes on three lines. There are nine points and nine lines in Figure 1.8, and these may be studied as a finite geometry with the familiar Euclidean properties no longer evident and with only the following axioms.

Axioms for Finite Geometry of Pappus

1. There exists at least one line.
2. Every line has exactly three points.
3. Not all points are on the same line.
4. There exists exactly one line through a point not on a line that is parallel to the given line.
5. If *P* is a point not on a line, there exists exactly one point *P'* on the line such that no line joins *P* and *P'*.
6. With the exception in Axiom 5, if *P* and *Q* are distinct points, then exactly one line contains both of them.

Study Figure 1.8 to see that each line of the finite geometry seems to have exactly two lines parallel to it. Of course, this observation does not constitute a proof. Axiom 4 is the familiar parallel axiom of Euclidean geometry, in a somewhat different setting.

The next theorem gives an additional property that can be proved from the axioms.

THEOREM 1.10. Each point in the geometry of Pappus lies on exactly three lines.

By Axioms 1 and 2, there exists a line with three points A, B, C on it (Figure 1.9). By Axiom 3, there is a fourth point (point X in Figure 1.9) not on this line. Consider the total number of lines on X, which represents any point of the geometry. By Axiom 5, X lies on lines meeting two of the points on the given line, say B and C. By Axiom 4, there is exactly one line through X parallel to $\overleftrightarrow{BC}$, so that there are at least three lines on X. But there cannot be a fourth line through X. By Axiom 5, there is no line connecting X and A, and by Axiom 4, there is no other line through X not meeting $\overleftrightarrow{BC}$.

FIGURE 1.9

Although the discussion of the geometry of Pappus began with the completed figure, it can be proved from the axioms alone (see Exercise Set 1.5) that the geometry of Pappus does in fact consist of exactly nine points and nine lines. Other versions of the axioms can be found using substitute words for points and lines, but it is also instructive to arrange the information in the form of a table. In Table 1.2, the notation from Figure 1.8 is used, with each vertical column representing a line of the geometry.

It is important to understand that a table such as Table 1.2 could be used, rather than a set of axioms, to give the initial representation of a geometry. Note, however, that in this case most proofs are very simple, since they depend just on checking all possible

TABLE 1.2

A	B'	D	A	A	F	F	B	C
B	A'	E	D	B'	C	B	A'	D
C	C'	F	C'	E	B'	C'	E	A'

cases in the table. For example, the table can be checked directly to see that each point lies on exactly three lines.

EXERCISES   1.5

1. Rewrite the set of axioms for the geometry of Fano, using *book* for *point* and *library* for *line*.

2. Which axioms in the geometry of Fano are also true statements in Euclidean geometry?

3. Using Figure 1.6c, name all the triangles in the geometry of Fano having point 4 as one vertex.

4. Using Figure 1.6c, if parallel lines are defined as those that intersect on line 1, 2, 3, name all the pairs of parallel lines in the geometry (do not use 1, 2, 3 as a member of the pairs).

5. For Fano's geometry, prove that each point lies on exactly three lines.

6. Rewrite the set of axioms for the geometry of Pappus, using the words *tree* for *point* and *row* for *line*.

7. Which axioms in the geometry of Pappus are also true statements in Euclidean geometry?

8. For each point in the geometry of Pappus, how many other points in the geometry do not lie on a line through the given point?

9. Prove that there are at least two lines in the geometry of Pappus parallel to a given line.

10. Prove that the geometry of Pappus contains exactly nine points.

11. Prove that the geometry of Pappus contains exactly nine lines.

1.6   OTHER FINITE GEOMETRIES

A good example of a modern finite geometry that is actually a study of a famous set of points from Euclidean geometry is the *finite geometry of Desargues.* Several concepts must be introduced so that they can be used in the study of this new geometry. Triangles *ABC* and *A'B'C'* of Figure 1.10 are *perspective from point P.* This means that point *P* is the common point (the point of concurrency) of the three lines joining corresponding vertices. Thus, *P* lies on $\overleftrightarrow{AA'}$, $\overleftrightarrow{BB'}$ and $\overleftrightarrow{CC'}$. According to Desargues' theorem, studied in more detail in Chapter 7, two triangles perspective from a point are also *perspective from a line.* If triangles are perspective from a line, corresponding sides of the triangles meet at points on this line. The line of perspectivity in Figure 1.10 contains points *R, S, T.* For example, corresponding sides $\overleftrightarrow{AB}$ and $\overleftrightarrow{A'B'}$ meet at *T,* a point on this line.

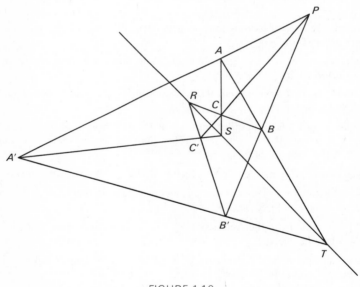

FIGURE 1.10

A further study of Figure 1.10 shows a total of ten labeled points on ten lines, with three points on each line and three lines on each point. These ten points and ten lines are the elements in the finite geometry of Desargues. One further concept is necessary to understand the axioms; a method of setting up a correspondence between points and lines is needed. If a point is the point of perspectivity for two triangles, and if a line is the line of perspectivity for the same two triangles, then the point is called the *pole* of the line and the line is called the *polar* of the point. In the finite geometry of Desargues, no line joins a pole and a point on a polar. This fact leads to the formal definitions, which do not use the concepts of point or line of perspectivity.

The line *l* in the finite geometry of Desargues is a polar of the point *P* if there is no line connecting *P* and a point on *l*.

The point *P* in the finite geometry of Desargues is a pole of the line *l* if there is no point common to *l* and any line on *P*.

The intuitive development has led to the axioms, but now the axioms become the beginning ideas; nothing can be proved from Figure 1.10—proof must come from the axioms themselves.

See if you can identify various poles and polars in Figure 1.10. For example, if point *T* is taken as the pole, where are the polars? This question can be answered by considering the lines through *T* and noting that there is exactly one line in the figure that has no points in common with these lines. That line is $\overleftrightarrow{PC'}$, hence $\overleftrightarrow{PC'}$ is the polar.

Axioms for the Finite Geometry of Desargues

1. There exists at least one point.
2. Each point has at least one polar.
3. Every line has at most one pole.
4. Two distinct points are on at most one line.
5. Every line has at least three distinct points on it.
6. If a line does not contain a certain point, then there is a point on both the line and any polar of the point.

The exact number of poles for a polar and the exact number of polars for a pole are not stated explicitly in the axioms. Study Figure 1.11. Assume that line *p* is a polar of *P*, since Axiom 2

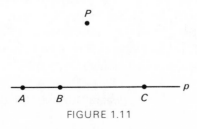

FIGURE 1.11

guarantees the existence of at least one. By definition, no line through *P* contains any one of the points (at least three) on *p*. But this also says that there is no point common to *p* and any line on *P*, so that *P* satisfies the definition of a pole of *p*. This information, along with Axioms 2 and 3, leads to the following theorems.

THEOREM 1.11. Every line of the geometry of Desargues has exactly one pole.

THEOREM 1.12. Every point of the geometry of Desargues has exactly one polar.

Parallel lines exist in the geometry of Desargues, but their properties are different from ordinary Euclidean parallels. For example, note in Figure 1.10 that three different lines can be drawn parallel to line *R, C, B* through point *A'* but that only one line can be drawn parallel to line *A, B, T*.

It is worthwhile to study the axioms and the drawings for a geometry to make additional conjectures. Proofs of additional theorems for the finite geometry of Desargues appear as exercises.

Several other finite geometries are discussed briefly in the rest of this section so that the interested reader can develop some of their theorems on his own. The starting place for each could be a set of axioms, although that approach may not always be the most productive when intuition is involved.

Fano's finite geometry has the special property of being *self-dual*. In other words, the plane dual of each true statement is a true statement for the geometry. Actually, Fano discussed a particular set of finite geometries, each of which was self-dual. The general symbol for geometries of this special type is $PG(n,q)$. The letters *PG* stand for projective geometry, to be discussed in Chapter 7. The letter *n* is the number of dimensions, and *q* is the positive integral power of a prime number. The geometry has $q + 1$ points on each line. Thus, Fano's geometry of the plane is $PG(2,2)$, since there are three points on each line. The theory of projective geometry can be used to establish the following useful theorem, assumed here.

THEOREM 1.13. The general formula for the total number of points in $PG(n,q)$ is

$$\frac{q^{n+1} - 1}{q - 1}.$$

For the geometry of Fano,

$$\frac{q^{n+1} - 1}{q - 1} = \frac{2^3 - 1}{1} = 7.$$

If $q = 3$, then $PG(2,3)$ is a new finite geometry that is self-dual. From Theorem 1.13, the total number of points is 13. $PG(2,3)$ is a finite geometry of 13 points and 13 lines. This geometry has the same axioms as Fano's geometry, except that there are four points rather than three on every line.

A different finite geometry, no longer self-dual, can be obtained from Fano's geometry by a modification of the last axiom. The new geometry, called *Young's geometry*, has the first four axioms of Fano's geometry along with the following substitute axiom for Axiom 5.

5. There is exactly one line on a point and not on any point on a line not containing the point.

Axiom 5 shows that the ordinary Euclidean concept of parallel lines applies, since it means that exactly one parallel to a given line can be found passing through each point not on the given line. Young's geometry is a finite geometry of nine points and twelve lines.

The interested student can make up finite geometries of his own, although some of them may be of limited significance. Various examples

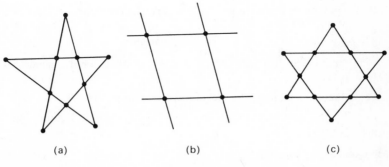

(a)   (b)   (c)

FIGURE 1.12

of simple drawings that can be used for a finite geometry are shown in Figure 1.12. A few of the interesting properties are mentioned in the exercises that follow.

EXERCISES   1.6

Exercises 1–8 concern the finite geometry of Desargues.

1. Prepare a table to represent the geometry of Desargues, using the points as named in Figure 1.10 and letting each column of the table represent a line in the geometry.

2. In Figure 1.10, name the pole of:

    a. $\overleftrightarrow{RB'}$ b. $\overleftrightarrow{AS}$

3. In Figure 1.10, name the polar of:

    a.   Point R b.   Point S

4. Using Figure 1.10, name the pair of triangles in the Euclidean figure perspective from point T and from $\overleftrightarrow{PC'}$.

5. Which axioms in the geometry of Desargues are also true statements in Euclidean geometry?

6. In Figure 1.10, name all the lines parallel to $\overleftrightarrow{AS}$.

7. Prove that if point P is on the polar of point Q, then point Q is on the polar of point P.

8. Prove that two lines parallel to the same line are not parallel to each other.

9. Prepare a table for the finite geometry $PG(2,3)$, using the numbers 1–13 for the points and showing the points on a line as one column of the table.

10. For $PG(2,3)$, there are how many lines on each point?

11. Which of these symbols represent self-dual geometries?

    a. $PG(2,4)$ b. $PG(2,5)$
    c. $PG(2,6)$ d. $PG(2,7)$

12. Find the number of points and lines, where possible, for the geometries of Exercise 11.

13. Prove that Young's geometry includes at least nine points.

14. For Young's geometry, prove that two lines parallel to a third line are parallel to each other.

For the geometry of Figure 1.12a:

15. Each line has how many points on it?

16. The geometry consists of how many points and how many lines?

17. Each line has how many other lines parallel to it?

For the geometry of Figure 1.12b:

18. Each line has how many points on it?

19. The geometry consists of how many points and how many lines?

20. Each line has how many other lines parallel to it?

For the geometry of Figure 1.12c:

21. The geometry consists of how many points and how many lines?

22. Each line has how many other lines parallel to it?

CHAPTER 2

GEOMETRIC TRANSFORMATIONS

2.1 INTRODUCTION TO TRANSFORMATIONS

In high school geometry, it is customary to speak of rotation and translation. Some triangles are proved congruent, and others are proved similar. In other types of geometry to be studied, one talks of the ideas of inversion and projection. All of these ideas are related to a very basic concept in geometry, that of a *geometric transformation.*

In this chapter, geometric transformations are defined and illustrated. The major examples are the set of transformations of Euclidean geometry. In addition, the study of transformations of similarity and

33

finite sets of transformations contributes to increased understanding of the concept.

Before giving a precise definition of transformation, it is necessary to explain the idea of mapping.

DEFINITION. A *mapping* of a set A *onto* a set B is a pairing of elements of A and B so that each element of A is paired with *exactly one* element of B, and each element of B is paired with *at least one* element of A.

Figure 2.1 shows an example of a mapping whose ordered pairs are (a_1, b_1), (a_2, b_1), and (a_3, b_2). A mapping of set A onto set B may also be indicated by the notation $b = f(a)$. Here, an element b of B is the image of an element a of A under the mapping f. For a mapping of set A *into* set B, it is not required that each element of B be involved in the pairing. In this case, A could be mapped onto a proper subset of B. It is assumed that the reader has a knowledge of elementary concepts of set theory.

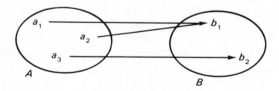

FIGURE 2.1

A special kind of onto mapping is of particular importance in mathematics; this is a one-to-one onto mapping, or a *transformation*.

DEFINITION. A transformation is a mapping f of A onto B such that each element of B is the image of exactly one element of A.

In other words, a one-to-one correspondence exists between the sets of elements of A and B. A simple example of a transformation is shown in Figure 2.2.

FIGURE 2.2

Recall that a *function* is a set of ordered pairs with no two different pairs having the same first element. It can be seen that the definition of mapping is equivalent to a typical definition of function, since in a mapping, having two pairs with the same first element would mean that an element of A is paired with more than one element of B. It is customary to use mapping rather than function in geometry, however, when the sets being considered are sets of points. One of the essential differences between a mapping and a transformation is that reversing the elements in the pairs of a mapping does not necessarily result in a mapping, while reversing the elements in the pairs of a transformation also results in a transformation. Make sure you understand why this is so.

Here are other examples of transformations to help make the concept clear.

a. A pairing of points on a number line indicated by $x \rightarrow 2x + 3$. For example, points with coordinates 1 and 5 or 2 and 7 would be paired. The arrow notation simply designates that the second element is the image of the first under the transformation. A summary of notation used in this text is found in Appendix 4.

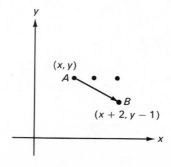

FIGURE 2.3

b. $(x, y) \rightarrow (x + 2, y - 1)$ is an example of a translation, a type of transformation that will be studied in more detail in a later section. As shown in Figure 2.3, each point in the plane is paired with a point two units to the right and one unit below the original point.

The concept of transformation is important because of the fact that sets of transformations can be used to classify geometries. In high school geometry, for example, the transformations allowed are rotation, translation, and sometimes similarity. Allowing other, more general types of transformations results in other modern geometries. In a particular geometry, the student studies *properties* of figures and their images under a set of transformations. *Invariant properties* are those that do not change. In the study of geometry through transformations, the student is asked to notice close relationships between modern algebra and modern geometry. Readers who have not yet studied abstract algebra will find necessary concepts explained in this text.

If f is a transformation from A onto B and g is a transformation from B onto C, the product $h = gf$ is defined as the transformation from A onto C such that $h(P) = g[f(P)]$ for each point of A. Note that the product gf is defined in such a way that the transformation on the right is performed first.

In Figure 2.4, the pairs in h are (a_1, c_3), (a_2, c_2), (a_3, c_1).

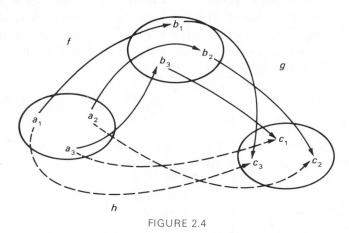

FIGURE 2.4

As an example of finding the product of two transformations, consider the two defined as follows for all real points (x, y).

$$f \quad (x, y) \to (x + 1, y + 1)$$

$$g \quad (x, y) \to (x + 2, y - 2)$$

In this example, note that the ordered pair has elements that are points expressed using an x and y coordinate in the Cartesian coordinate system of analytic geometry.

The results of f followed by g, (gf) for the sample points $(2, 2)$ and $(1, 5)$ are shown in Figure 2.5. The combined effect of gf is to cause (x, y) to have the image $(x + 3, y - 1)$.

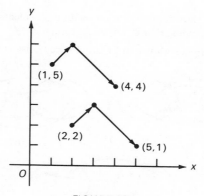

FIGURE 2.5

In special cases, the product of two transformations is the *identity transformation*, I, (a_1, a_1), $(a_2, a_2), \ldots, (a_n, a_n)$. For the identity transformation, each element is its own image. The identity transformation leaves each point fixed. In this case, if $I = gf$, as illustrated

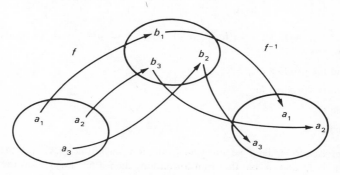

FIGURE 2.6

in Figure 2.6, then g is the *inverse transformation* of f, indicated by f^{-1}, so that $f^{-1}f = ff^{-1} = I$. Recall that the inverse of a transformation has the effect of "undoing" the original transformation, so that the product of a transformation and its inverse is the identity transformation. For example, if f is a transformation such that (x, y) has the image $(x + 4, y - 2)$, then f^{-1} is a transformation such that (x, y) has the image $(x - 4, y + 2)$.

EXERCISES   2.1

1.   Draw a diagram showing two sets, one with six elements and one with eight; then indicate a mapping that is not a transformation.

2.   Which of the mappings shown in Figure 2.7 are transformations?

(a)   (b)

(c)

FIGURE 2.7

3.   For the transformation indicated by $(x, y) \rightarrow (x + 5, y - 3)$, give the image of these points:

   a. $(2, 1)$ b. $(0, 0)$ c. $(-3, -2)$

For Exercises 4–7, let $f = \{(a, b), (c, d), (e, h)\}$ and $g = \{(b, i), (d, j), (h, k)\}$ be transformations.

4.   Find the product gf.

5.   Find the product fg.

6.   Find f^{-1}.

7.   Find g^{-1}.

For Exercises 8–11, let f be the transformation such that (x, y) has the image $(x - 5, y + 2)$ and g be the transformation such that (x, y) has the image $(x + 2, y - 3)$.

8.   Find the product fg.

9.   Find the product gf.

10. Find f^{-1}.

11. Find g^{-1}.

2.2   GROUPS OF TRANSFORMATIONS

In modern geometry, sets of transformations are ordinarily discussed rather than individual transformations.

DEFINITION. A *group of transformations* is a nonempty set S of transformations f such that:
   a. $f \in S$ implies that $f^{-1} \in S$,
   b. $f \in S$ and $g \in S$ imply that $fg \in S$,
   c. $f(gh) = (fg)h$.

The first of the properties above guarantees that the inverse for each transformation is also an element in the set of transformations. The second of the properties is the *closure* property. This means the product of any two transformations in the set is also a transformation in the set. Stated another way, the operation of multiplication can always be performed within a group of transformations without going outside the group. The last of the properties is the *associative* property. It is assumed here that, regardless of the grouping by parentheses, the definition of multiplication of transformations results in one transformation followed by another, and $f(gh)$ and $(fg)h$ imply the product of the same three transformations in the same order. For example, consider the three transformations which have been defined as follows:

$$f: x \to x + 3,$$

$$g: x \to x - 2,$$

$$h: x \to 2x,$$

$$f(gh): x \to (2x - 2) + 3 = 2x + 1,$$

$$(fg)h: x \to 2x + 1.$$

A fourth property that needs to be established for every group of transformations is the existence of the *identity element* as an element of the group. This is left as an exercise. In general, the commutative property *fg = gf* does not hold for transformations. For example, the finite group to be introduced shortly is not a commutative group. If the commutative property does hold for a group, that group is called a *commutative* or *Abelian* group (after the Norwegian mathematician N. H. Abel, 1802–1829).

Many of the groups of transformations in geometry are infinite groups—that is, groups with an infinite number of members. On the other hand, examples of finite groups will help in developing the concept of a group of transformations. It is important to remember that the elements of a group of transformations are transformations, not points.

Consider the set of all *symmetries of an equilateral triangle,* as shown in Figure 2.8. "Symmetries of the triangle" designates the reflections

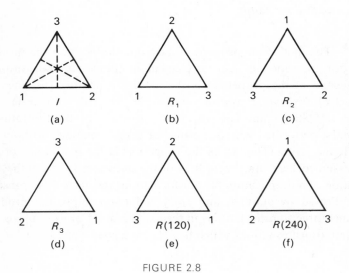

FIGURE 2.8

about the axes of symmetry or rotation such that the new figure coincides with the old. The result of each symmetry can be represented by a renaming of the three vertices. Check each of the figures in Figure 2.8 to see that it corresponds to the following definitions of the elements of the set of symmetries of an equilateral triangle:

R_1 Reflection about the axis through vertex 1.
R_2 Reflection about the axis through vertex 2.
R_3 Reflection about the axis through vertex 3.
$R(120)$ Rotation through an angle of 120° counterclockwise.
$R(240)$ Rotation through an angle of 240° counterclockwise.
I Identity.

To verify that the set of symmetries of an equilateral triangle constitutes a group, it is necessary to verify that $f \in S$ implies that $f^{-1} \in S$ and that $f \in S$ and $g \in S$ implies $fg \in S$. The inverse element for each transformation is listed below.

f	f^{-1}
R_1	R_1
R_2	R_2
R_3	R_3
$R(120)$	$R(240)$
$R(240)$	$R(120)$
I	I

The verification of both facts can be accomplished by completing a multiplication table for the elements. The entries in the table are found by performing the transformations in the order indicated. For example, Table 2.1 shows $R_2[R(120)] = R_3$. Recall that $R(120)$ comes first in this multiplication.

TABLE 2.1

First Transformation Performed

	I	R_1	R_2	R_3	$R(120)$	$R(240)$
I	I	R_1	R_2	R_3	$R(120)$	$R(240)$
R_1	R_1	I	$R(120)$	$R(240)$	R_2	R_3
R_2	R_2	$R(240)$	I	$R(120)$	R_3	R_1
R_3	R_3	$R(120)$	$R(240)$	I	R_1	R_2
$R(120)$	$R(120)$	R_3	R_1	R_2	$R(240)$	I
$R(240)$	$R(240)$	R_2	R_3	R_1	I	$R(120)$

Second Transformation Performed

Introduction of the *permutation group symbols* will illustrate a common agreement that ties the work on transformation groups more closely to that found in modern algebra texts. Two rows of numbers are used to define a transformation. The first row shows the original vertices and the second row shows the new position of those vertices.

$$R_1 \quad = \begin{pmatrix} 1, & 2, & 3 \\ 1, & 3, & 2 \end{pmatrix}$$

Explanation for R_1. Vertex 1 remains fixed, vertex 2 moves to the place where vertex 3 was originally, and vertex 3 moves to the place where vertex 2 was originally.

$$R_2 \quad = \begin{pmatrix} 1, & 2, & 3 \\ 3, & 2, & 1 \end{pmatrix}$$

$$R_3 \quad = \begin{pmatrix} 1, & 2, & 3 \\ 2, & 1, & 3 \end{pmatrix}$$

$$R(120) = \begin{pmatrix} 1, & 2, & 3 \\ 2, & 3, & 1 \end{pmatrix}$$

Explanation for $R(120)$. Vertex 1 moves to the place where vertex 2 was originally, vertex 2 moves to the place where 3 was originally, and vertex 3 moves to the place where 1 was originally.

$$R(240) = \begin{pmatrix} 1, & 2, & 3 \\ 3, & 1, & 2 \end{pmatrix}$$

There are five other equivalent forms for each of the permutation group symbols. This is true because the elements on the first row can be arranged in any of six ways. For example, two other equivalent forms for R_1 are

$$\begin{pmatrix} 3, & 2, & 1 \\ 2, & 3, & 1 \end{pmatrix} \quad \text{and} \quad \begin{pmatrix} 3, & 1, & 2 \\ 2, & 1, & 3 \end{pmatrix}.$$

Either the permutation symbols, or a series of pictures, can be used to check the entries in the multiplication table. Three examples follow to illustrate both methods.

EXAMPLE 1

$$R_1 R_2 = \begin{pmatrix} 3, & 2, & 1 \\ 2, & 3, & 1 \end{pmatrix}\begin{pmatrix} 1, & 2, & 3 \\ 3, & 2, & 1 \end{pmatrix} = \begin{pmatrix} 1, & 2, & 3 \\ 2, & 3, & 1 \end{pmatrix} = R(120)$$

The notation means that R_2 is first. This is indicated by

$$\begin{pmatrix} 1, & 2, & 3 \\ 3, & 2, & 1 \end{pmatrix}.$$

Then the result on the bottom line becomes the top line of one of the equivalent forms for R_1,

$$\begin{pmatrix} 3, & 2, & 1 \\ 2, & 3, & 1 \end{pmatrix}.$$

The product has the top row from R_2 and the bottom row from the form of R_1, and can be identified as the definition of $R(120)$.

Finding the product of the two transformations can also be illustrated as in Figure 2.9.

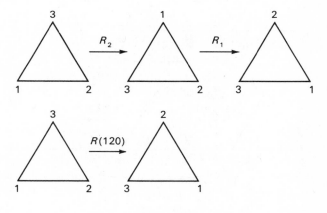

FIGURE 2.9

Note the important convention observed in the first row of Figure 2.9. The second transformation is performed using the original positions for the vertices. Thus, the vertex that was where vertex 3 was originally, (1), is moved to where vertex 2 was originally. The vertex that was where vertex 2 was originally, (2), is moved to where vertex 3 was originally, and the vertex that was where vertex 1 was originally, (3), remains unchanged. Because of this somewhat tedious interpretation, the reader will probably find the use of permutation symbols a valuable shortcut.

EXAMPLE 2

$$R(120)R_2 = \begin{pmatrix} 3, & 2, & 1 \\ 1, & 3, & 2 \end{pmatrix}\begin{pmatrix} 1, & 2, & 3 \\ 3, & 2, & 1 \end{pmatrix} = \begin{pmatrix} 1, & 2, & 3 \\ 1, & 3, & 2 \end{pmatrix} = R_1$$

EXAMPLE 3

$$R(240)R(120) = \begin{pmatrix} 2, & 3, & 1 \\ 1, & 2, & 3 \end{pmatrix}\begin{pmatrix} 1, & 2, & 3 \\ 2, & 3, & 1 \end{pmatrix} = \begin{pmatrix} 1, & 2, & 3 \\ 1, & 2, & 3 \end{pmatrix} = I$$

Examples 2 and 3 are illustrated in Figure 2.10.

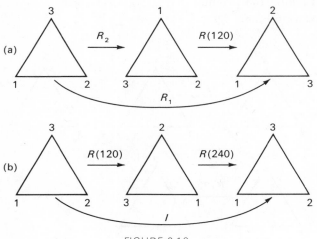

FIGURE 2.10

The conclusion from the discussion of the symmetries of the equilateral triangle is stated as a theorem.

THEOREM 2.1. The set of symmetries of the equilateral triangle is a group of transformations.

The group of symmetries of an equilateral triangle is not a commutative group. As a specific example, $R_1 R(240) = R_3$, but $R(240)R_1 = R_2$. Note that a consequence of this lack of commutativity is that, when the elements are listed in the same horizontal and vertical order, the multiplication table is not symmetric about the diagonal from upper left to lower right.

Within a group of transformations may be subgroups of transformations. A *subgroup* is a subset of a group that is itself a group. For example, one such subgroup of the symmetries of an equilateral triangle has elements I, $R(120)$, and $R(240)$. Verify from the multiplication table that the products are elements of a subgroup. Furthermore, observe that this subgroup is a commutative group.

Other examples of finite groups of transformations are the symmetries of a square and the symmetries of an isosceles triangle. These are explored in the following set of exercises.

EXERCISES   2.2

1. Prove that a group of transformations includes the identity element.

2. Write an explanation for the permutation group symbols for R_2, R_3, and $R(240)$.

For Exercises 3–6, find the products from the multiplication table for the symmetries of an equilateral triangle, then use the permutation notation to verify each answer.

3. $R(240)R_3$ 4. $R_3 R(240)$

5. $R_1 R_3$ 6. $R_3 R_1$

7. Is the set $\{I, R_1, R_2, R_3\}$ of symmetries for the equilateral triangle a subgroup?

8. Prepare a multiplication table for the symmetries of an isosceles triangle.

9. Verify the fact that the symmetries of an isosceles triangle form a group.

10. List all the subgroups of the symmetries of an isosceles triangle.

11. Prepare a multiplication table for the symmetries of a square.

12. Verify the fact that the symmetries of a square form a group.

13. List all the subgroups of the symmetries of a square.

2.3   EUCLIDEAN MOTIONS OF A PLANE

The examples of transformations that are most familiar are those used in high school geometry. Unlike the finite groups of transformations studied in the previous section, the sets of transformations in this section have an infinite number of members. The detailed study in this section and the next two involves looking at Euclidean geometry from a point of view different from the typical secondary school approach.

The essential characterization of the transformations of Euclidean geometry is that *distance* is preserved. That is, distance must be an invariant property.

DEFINITION. A transformation f is an *isometry* of A onto B if it preserves distances. For any two points P_1, P_2 of A,

$$|f(P_1)f(P_2)| = |P_1 P_2|,$$

where $|P_1 P_2|$ denotes the distance between P_1 and P_2.

FIGURE 2.11

This idea is illustrated in Figure 2.11. The symbol $|P_1 P_2|$ means the absolute value of the distance, so that only positive (undirected) distances need be considered here. Sometimes, distance is an undefined term in synthetic geometry, but in analytic geometry the distance d between points (x_1, y_1) and (x_2, y_2) is defined by

$$d = \sqrt{(x_2 - x_1)^2 + (y_2 - y_1)^2}.$$

The close relationship between the new concept of isometry and a more familiar concept is apparent with the added statement that, if an isometry exists between two sets of points, the sets are *isometric*, or *congruent*.

In ordinary Euclidean geometry, the isometry that is studied is an isometry of a set of points onto itself. In other words, the image of each point of the plane is another point in the same plane. Isometries of this kind are called *motions*. Euclidean geometry involves the study of motions of the real plane.

In 1872, Felix Klein classified geometries by applying this definition: A geometry is the study of invariant properties of a set of points under a group of transformations. Thus, Euclidean geometry is the study of invariant properties, such as angle measure and area, of sets of points under the group of Euclidean transformations.

The reader should ask himself, before continuing, what are the different kinds of Euclidean motions. The same question can be stated differently by asking what types of transformations can be applied to a set of points in the plane so the distance between any two points is always preserved. This question is answered in this section, but the answer for the same question applied to analytic geometry is reserved for the next section.

The first type of plane motion is a *translation*. Intuitively, a translation sets up a correspondence between points and their image points so that each image is the same distance in the same direction from the original point. Figure 2.12a shows a translation of segment $\overline{AB}$ in a direction represented by the vector AA'. Figure 2.12b shows a translation of triangle ABC in a direction represented by the vector BB'. Check each of these observations about a translation.

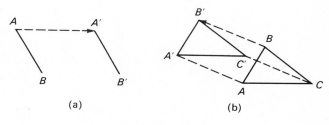

(a)   (b)

FIGURE 2.12

a. A segment is translated into a parallel segment.
b. All of the vectors connecting corresponding points are equal.
c. The inverse of a translation is another translation the same distance in the opposite direction.
d. As illustrated in Figure 2.13, the product of two translations is a translation. The vector for the product is the sum of the vectors for the two translations.
e. The set of all translations forms a group.

Additional information about translations can be found in the next sections, where the discussion includes the use of coordinates.

The second basic type of motion of the plane is a *rotation.* The symbol $R(O, \alpha)$ indicates a rotation through an angle of α about the point $O,$ as in Figure 2.14. By convention, a counterclockwise rotation is associated with a positive angle.

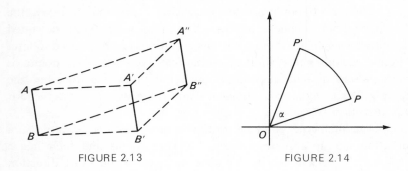

FIGURE 2.13   FIGURE 2.14

Figure 2.15 shows rotation of a segment and a triangle about a point. In both cases, the angle of rotation is $\angle AOA'$. Check these observations about the motion of rotation.

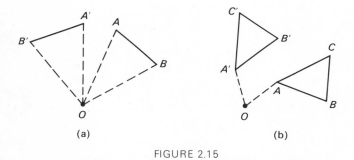

(a)   (b)

FIGURE 2.15

  a. A segment is not necessarily parallel to its image.
  b. The inverse of a rotation is a rotation about the same point and with the same angle of rotation, but measured in the opposite sense.
  c. The product of two rotations about the same point is another rotation about that point.
  d. The set of all rotations about a fixed point is a group of transformations.

The set of all translations and rotations is called the set of *rigid motions* or *displacements*. Using a cardboard model of a triangular region makes it possible to illustrate rigid motions by sliding the model from one position to another in a rigid way without changing its size or shape.

The third basic example of motion of a plane is a *reflection*. A reflection R_l utilizes a fixed line l, as in Figure 2.16. A point on l is its own image. Any other point P is mapped into a point P' such that l is the perpendicular bisector of $\overline{PP'}$. (The notation $\overline{PP'}$ means the segment with P and P' as endpoints.) As the name *reflection* implies, a set of points and their images are reflections of each other; it is as if line l were a mirror. Note that a set of points having l as a line of symmetry is mapped onto itself by a reflection about l. One half of the figure is the image of the other half.

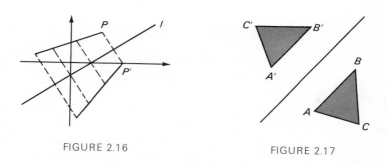

FIGURE 2.16   FIGURE 2.17

Check these additional observations about reflections.
a. In general, a segment is not parallel to its image.
b. The inverse of a reflection is the same reflection.
c. The product of two reflections about a line is not a reflection.
d. As illustrated in Figure 2.17, a reflection cannot be considered a sliding in the plane. It is necessary to flip over the
   cardboard model of the triangular region to have it correspond
   to the image.

The three transformations given so far, translation, rotation, and reflection, may be considered the three basic motions of the plane. But this set is not closed, and in fact one more transformation is necessary for closure. What this means is that some products of two of these transformations are neither a translation, a rotation, nor a reflection, but yet another type of transformation. Specifically, the product of a reflection

and one of the other two transformations may not be an element of
the set.

The fourth type of motion of a plane is a *glide reflection*.
A glide reflection is the product of a reflection followed by a translation
parallel to the fixed line of reflection, as in Figure 2.18, in which P is
mapped into P''.

FIGURE 2.18

The exercises include questions that will require experimentation
with drawings or cardboard models to develop additional information
about the glide reflection.

One geometric application of various motions of the plane is in
simple geometric designs, such as those shown in Figure 2.19, that
form plane-filling repeated patterns. The Dutch artist M. C. Escher has
extended this idea to very intricate plane-filling patterns involving such
figures as fish, birds, horsemen, and reptiles. The interested reader will
enjoy studying these works of art in *The Graphic Work of M. C. Escher*.

(a)   (b)

FIGURE 2.19

The somewhat intuitive introduction to plane motions in this
section is followed by two sections developing the concepts in greater

detail. Section 2.4 develops the material using analytic geometry, and Section 2.5 discusses the set of all motions of the plane from a more general point of view.

EXERCISES   2.3

1. For a translation, is the measure of the angle between two rays an invariant?

2. Could a translation be its own inverse?

3. Explain what is meant by the identity translation.

4. Explain what is meant by the identity rotation.

5. Explain how a segment and its image might be parallel for a rotation other than the identity.

6. Describe the inverse of a rotation using a positive direction instead of a negative one.

7. Give an example in which a segment and its image are parallel under reflection.

8. Describe the inverse of a glide reflection.

9. Are a segment and its image parallel under a glide reflection?

10. Can a model for a triangular region be made to coincide with its image under a glide reflection by sliding?

2.4   SETS OF EQUATIONS FOR MOTIONS OF THE PLANE

In this section, analytic geometry is utilized to approach the study of motions of the plane as sets of linear equations. If A is any

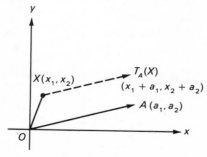

FIGURE 2.20

point of the plane, the transformation T_A such that $T_A(X) = X + A$ is a translation in the direction of OA. In Figure 2.20, the point $T_A(X)$ is the image of point X under the translation. Its coordinates are the sum of the coordinates of points X and A, $(x_1 + a_1, x_2 + a_2)$. Thus, the notation $X + A$ is used to indicate the addition of the corresponding coordinates of points X and A.

The concept of sets of equations for a translation should be familiar from elementary analytic geometry. In Figure 2.21, for example,

FIGURE 2.21

if $P(1,2)$ has the image $P'(3,3)$, then the translation may be represented by the equations

$$x' = x + 2,$$
$$y' = y + 1.$$

Also, using the notation of addition of coordinates of points, $(1, 2) + (2, 1) = (3, 3)$, hence $A = (2, 1)$. Note of course that $(2, 1)$ also indicates the vector PP' in the diagram.

DEFINITION. A translation is a transformation with equations of the form

$$x' = x + a,$$
$$y' = y + b,$$

for a and b any real numbers. The inverse of this transformation has the equations

$$x = x' - a,$$
$$y = y' - b.$$

EXAMPLE. Find the image of $(5, 2)$ under the translation with equations $x' = x + 7$, $y' = y + 3$. The solution is $(12, 5)$.

In typical analytic geometry textbooks, it is shown that the defining equations for a rotation about the origin are as follows.

DEFINITION. A rotation about the origin is a transformation with equations of the form

$$x' = x \cos \alpha - y \sin \alpha,$$
$$y' = x \sin \alpha + y \cos \alpha.$$

EXAMPLE. Find the image of $P(2, 3)$ under a rotation of $60°$ about the origin.

The equations are

$$x' = x \cos 60° - y \sin 60°,$$
$$y' = x \sin 60° + y \cos 60°.$$

Then

$$x' = (2)\left(\frac{1}{2}\right) - (3)\left(\frac{\sqrt{3}}{2}\right),$$

$$y' = (2)\left(\frac{\sqrt{3}}{2}\right) + (3)\left(\frac{1}{2}\right).$$

The coordinates of P' are

$$\left(1 - \frac{3}{2}\sqrt{3}, \sqrt{3} + \frac{3}{2}\right).$$

The equations of rotation about a point (h, k) are

$$x' - h = (x - h)\cos\alpha - (y - k)\sin\alpha,$$
$$y' - k = (x - h)\sin\alpha + (y - k)\cos\alpha.$$

Note how this transformation may be considered the product of a rotation and a translation.

The sets of equations for the reflection transformation are not as familiar as those for the transformations of translation and rotation. For the special case of reflection about the x-axis, the equations are

$$x' = x,$$
$$y' = -y.$$

The forms are also simple when the reflection is about the y-axis.

A reflection about the line $y = mx$, where $m = \tan\theta$ as shown in Figure 2.22, is a reflection about the x-axis, combined with a

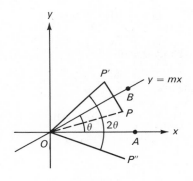

FIGURE 2.22

rotation through an angle of 2θ about the origin. This can be seen from Figure 2.22 because of the congruences $\angle POB \cong \angle P'OB$ and $\angle POA \cong \angle P''OA$. The equations for this reflection are

$$x' = x\cos 2\theta + y\sin 2\theta,$$
$$y' = x\sin 2\theta - y\cos 2\theta,$$

obtained by writing the equations for a rotation of 2θ about the origin, then substituting $-y$ for y.

EXAMPLE. Find the image of $(2, 3)$ under a reflection about the line $y = x$. Since $\cos 2\theta = 0$ and $\sin 2\theta = 1$,

$$x' = y,$$
$$y' = x,$$

hence,

$$x' = 3,$$
$$y' = 2,$$

and the point is $(3, 2)$, as should be expected.

The general equations for reflection about a line $y = mx + b$ can be obtained from the special case of reflection about a line through the origin by a translation through a distance twice the perpendicular distance d from $y = mx$ to $y = mx + b$. The reason for twice the distance is explained in the discussion of Theorem 2.5 in the next

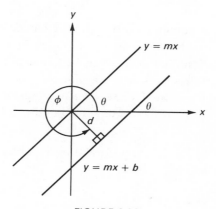

FIGURE 2.23

section, although the answer can also be determined here if desired. Since the translation is in the direction indicated by ϕ in Figure 2.23, the equations of the translation are

$$x' = x + 2d \cos \phi,$$
$$y' = y + 2d \sin \phi,$$

and the desired equations for the general reflection are

$$x' = x \cos 2\theta + y \sin 2\theta + 2d \cos \phi,$$
$$y' = x \sin 2\theta - y \cos 2\theta + 2d \sin \phi.$$

EXAMPLE. Find the image of point $(2, 6)$ under a reflection about the line such that $\theta = 30°$, $\phi = 300°$, and $d = 2$.

$$x' = x \cos 60° + y \sin 60° + 4 \cos 300°$$
$$y' = x \sin 60° - y \cos 60° + 4 \sin 300°$$
$$x' = (1/2)x + (\sqrt{3}/2)y + 2$$
$$y' = (\sqrt{3}/2)x - (1/2)y - 2\sqrt{3}$$

For $x = 2$, $y = 6$,

$$x' = 1 + 3\sqrt{3} + 2 = 3 + 3\sqrt{3},$$
$$y' = \sqrt{3} - 3 - 2\sqrt{3} = -3 - \sqrt{3}.$$

The general equations for a glide reflection are derived from those for a reflection and those for a translation in a direction parallel to the line of reflection. The images can be found using the two sets of equations given previously for these transformations.

A study of the form of the sets of equations for the four motions of the plane shows they may all be described in the same general way.

THEOREM 2.2. If a transformation is a plane motion, then it has equations of the form

$$x' = ax + by + c,$$
$$y' = \pm(-bx + ay) + d,$$

for $a, b, c, d \in R$ and $a^2 + b^2 = 1$.

The *converse* of this theorem is also true, although the proof is not given here. The converse of a theorem stated as an implication is obtained by interchanging the *if* and *then* parts of the statement. The converse is:

THEOREM 2.3. If a transformation has equations of the form

$$x' = ax + by + c,$$
$$y' = \pm(-bx + ay) + d,$$

for $a, b, c, d \in R$ and $a^2 + b^2 = 1$, then it is a plane motion.

The equations for the inverse of a transformation can be found by solving the set of equations for x and y. This was illustrated previously for a translation. The following example is for a reflection.

EXAMPLE. Find the inverse of the transformation with equations

$$x' = x \cos 30° + y \sin 30°,$$
$$y' = x \sin 30° - y \cos 30°.$$

Multiply the members of the first equation by $\sin 30°$ and the members of the second by $\cos 30°$.

$$x' \sin 30° = x \sin 30° \cos 30° + y \sin^2 30°$$
$$\underline{y' \cos 30° = x \sin 30° \cos 30° - y \cos^2 30°}$$

$$x' \sin 30° - y' \cos 30° = y$$

Now multiply both members of the original first equation by $\cos 30°$ and the members of the second by $\sin 30°$.

$$x' \cos 30° = x \cos^2 30° + y \sin 30° \cos 30°$$
$$\underline{y' \sin 30° = x \sin^2 30° - y \sin 30° \cos 30°}$$

$$x' \cos 30° + y' \sin 30° = x$$

As should be expected, a change of variables will make obvious the fact that the inverse of a reflection is the original transformation itself.

EXERCISES 2.4

1. Let the vector for a translation be $(3, -5)$ and give the image for each point under the translation.

   a. $(0,0)$ b. $(5,7)$
   c. $(-1, -8)$ d. $(-2, -\frac{3}{2})$

2. In the translation with vector $(5,9)$ what points have these points as images?

   a. $(0,0)$ b. $(5,9)$

3. What is the image of the point $(3,4)$ under a rotation of $45°$ about the origin?

4. What point has the image $(-1,2)$ under a rotation of $45°$ about the origin?

5. Derive the equations for the inverse of a transformation of rotation about a point (h, k).

6. Find the image of $(3,7)$ under a reflection about the line $y = (\sqrt{3}/3)x$.

7. Find the image of $(3,7)$ under a reflection about the line with $\theta = 60°$, $\phi = 330°$, and $d = 5$.

8. What is the image of $(5,7)$ under the glide reflection consisting of a reflection about the y-axis followed by a translation of three units in a positive direction parallel to the y-axis?

For Exercises 9–12, find the inverse transformation for the transformation whose equations are given:

9. $x' = x - 5$
   $y' = y + 2$

10. $x' = x \cos 30° - y \sin 30°$
    $y' = x \sin 30° + y \cos 30°$

11. $x' = x$
    $y' = -y$

12. $x' = x \cos 60° + y \sin 60°$
    $y' = x \sin 60° - y \cos 60°$

2.5 PROPERTIES OF THE EUCLIDEAN GROUP

The set of Euclidean motions of a plane, introduced in Sections 2.3 and 2.4, consists of translations, rotations, reflections, and glide reflections. In this section, additional properties of motions are considered.

THEOREM 2.4. The four basic Euclidean motions of the plane constitute a group of transformations.

The proof of Theorem 2.4 consists of showing that there exists an inverse for each element and that the product of any two elements is another element of the set.

It is easy enough to verify that each element of the four types of motions named has an inverse element. For example, the inverse of a translation is another translation in the opposite direction. The check for closure with respect to multiplication is somewhat more complex. A proof using analytic geometry could be given, using Theorems 2.2 and 2.3 of the previous section. The synthetic proof is normally facilitated by showing that motions of the plane may be discussed as products of reflections.

THEOREM 2.5. Every plane motion is the product of three or fewer reflections and conversely.

The theorem is proved in two parts. The first part is for translations and rotations, while the second part is for reflections and glide reflections.

1. Every translation and rotation is the product of two reflections and conversely.

The two cases depend on whether the two lines of reflection intersect or are parallel.

    a. If l_1 and l_2 are any two lines of the plane intersecting at O, and if the angle from l_1 to l_2 is θ, then $R(O, 2\theta) = R_{l_1} R_{l_2}$.

In Figure 2.24, the rotation that transforms $\overline{PQ}$ into $\overline{P''Q''}$ can be considered the product of two reflections about l_1 and l_2.

    b. If l_1 and l_2 are parallel lines, and if they are perpendicular to a translation vector OP and at a distance $|OP|/2$ apart, then $T_p = R_{l_2} R_{l_1}$.

FIGURE 2.24

In Figure 2.25, the motion that transforms points A and B into A'' and B'' may be considered the product of two reflections about l_1 and l_2.

FIGURE 2.25

2. Every reflection or glide reflection is the product of three or fewer reflections. This statement is trivial for a reflection itself. Since a glide reflection is the product of a reflection and a translation, and since a translation is the product of two reflections, a glide reflection is the product of three reflections.

The proof of the converse of Theorem 2.5 is not completed. The converse states that the product of three or fewer reflections is a motion of the plane. The type of motion depends on the relative position of the lines of reflection. The possibilities are summarized in Table 2.2.

TABLE 2.2

Product of Two Reflections	A. If the two lines of reflection are parallel, then the motion is a translation. B. If the two lines of reflection are nonparallel, then the motion is a rotation.
Product of Three Reflections	A. If two of the lines of reflection coincide, then the motion is a reflection. B. If the three lines of reflection are parallel, then the motion is a reflection. C. If two lines of reflection intersect at a point on the third, then the motion is a reflection. D. If two lines of reflection intersect at a point not on the third, then the motion is a glide reflection.

In Section 2.3, experimentation led to the fact that some motions could be represented by sliding a cardboard model in the plane, while others could not. This matter is now reconsidered in the light of Theorem 2.5.

DEFINITION. A plane motion is a *direct* motion if it is the product of an even number of reflections. It is an *opposite* motion if it is the product of an odd number of reflections.

According to this definition, rotations and translations are direct motions, whereas reflections and glide reflections are opposite motions. Intuitively, the difference between direct and opposite motions can be seen by considering a triangular region cut out of cardboard, such as $\triangle ABC$ in Figure 2.26. This piece of cardboard can be moved in the plane to represent any direct motion. For example, it can be translated to the position of $\triangle A'B'C'$. On the other hand, the paper must be turned over if the movement is to represent an opposite motion. It is necessary, for example, to turn the cardboard triangular region over for it to coincide with $\triangle A''B''C''$.

FIGURE 2.26

The general equations for motions include a $\pm$ symbol. The plus sign indicates a direct motion, and the negative sign indicates an opposite motion. This may be verified by a check of each individual type of equation for a motion.

The product of direct and opposite motions is summarized in Table 2.3.

TABLE 2.3

	Direct	*Opposite*
Direct	Direct	Opposite
Opposite	Opposite	Direct

As one with sufficient algebraic background might suspect, the system of direct and opposite motions under multiplication is isomorphic to the system of even and odd numbers under addition (or the integers modulo 2). This can be seen by comparing Table 2.3 with the addition tables derived for the other two systems mentioned.

The use of Theorem 2.5 and the concepts of direct and opposite motions finally makes it possible to complete a table showing closure for Theorem 2.4. Since each entry in Table 2.4 shows two possibilities, more information about the beginning motions is necessary in order to classify the product as a unique type of motion.

TABLE 2.4

	Rotation	*Translation*	*Reflection*	*Glide Reflection*
Rotation	rotation or translation	rotation or translation	glide reflection or reflection	glide reflection or reflection
Translation	rotation or translation	rotation or translation	glide reflection or reflection	glide reflection or reflection
Reflection	glide reflection or reflection	glide reflection or reflection	rotation or translation	rotation or translation
Glide Reflection	glide reflection or reflection	glide reflection or reflection	rotation or translation	rotation or translation

Recall that a geometry consists of the study of properties invariant under a group of transformations. The group of plane motions, for example, are the transformations of ordinary Euclidean geometry. Basic to the investigation of other invariant properties is the question of invariant points. Sometimes a point and its image are identical, and this result affects the study of any other invariant properties. The effect of a motion on the points of a plane is to change many points and possibly to leave some invariant.

For each type of motion, the possibilities for numbers of fixed points without being the identity transformation are summarized in the form of Table 2.5.

TABLE 2.5

Motion	Possibilities for Invariant Points
translation	none
rotation	one point
reflection	one line
glide reflection	none

At most, one line of the plane may remain invariant in a motion that is not the identity. In the light of this discovery, it is significant to ask about the minimum number of points and their images needed to uniquely determine a plane motion other than the identity. Proof of the following theorem will establish that the answer is three noncollinear points.

THEOREM 2.6. A unique motion of the plane is completely determined by an isometry of one triangle onto a second.

Let $\triangle ABC \cong \triangle A'B'C'$ in Figure 2.27. It is necessary to show there is exactly one motion of the plane agreeing with this isometry.

FIGURE 2.27

Let P be any point of the plane other than A, B, or C. If the isometry is direct, then P' is uniquely determined by constructing $\angle B'A'P' \cong \angle BAP$ and $A'P' = AP$. (The symbol $A'P'$ means the distance from A' to P', rather than the segment itself.) If it is opposite, the angle is constructed in the opposite sense.

Let P_1 and P_2 be two points of the plane in Figure 2.28. $\triangle AP_1 P_2 \cong \triangle A'P_1'P_2'$, since $AP_1 = A'P_1'$, $AP_2 = A'P_2'$, and $\angle P_1 AP_2 \cong \angle P_1'A'P_2'$. Thus, $P_1 P_2 = P_1'P_2'$, so that a motion is determined because distance between two points is preserved.

FIGURE 2.28

Since, for any given point P, there is only one point of the plane P' such that $PA = P'A'$, $PB = P'B'$, and $PC = P'C'$, P' is the image of P, and the motion is the only one for which $\triangle ABC$ has as its image $\triangle A'B'C'$.

The properties studied in elementary geometry, such as congruence of segments, triangles, and angles, area of regions, and intersection of lines, are all properties of sets of points that are invariant under the group of motions.

The proof of the invariance of a property under a group of transformations can be either synthetic or analytic. Both types of proofs are illustrated in examples.

EXAMPLE. Prove that a segment and its image are parallel under a translation.

In Figure 2.29, let $\overline{AB}$ be the segment and $\overline{A'B'}$ its image.

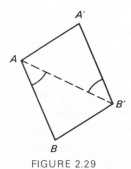

FIGURE 2.29

$\triangle AA'B' \cong \triangle B'BA$, because pairs of corresponding sides are congruent. Thus, $\angle AB'A' \cong \angle B'AB$, and $\overline{AB}$ and $\overline{A'B'}$ are parallel.

EXAMPLE. Prove analytically that a line and its image are parallel under a translation.

Let the equation of the line be $ax + by + c = 0$ and the equations of the translation be $x' = x + d$, $y' = y + e$. The equations of the inverse are $x = x' - d$, $y = y' - e$, so that the image of the line is $a(x' - d) + b(y' - e) + c = 0$, or $ax' + by' + (-ad - be + c) = 0$. This equation and the original equation have the same slope, so that the lines are parallel.

The location of the points is virtually the only property not an invariant in a motion. The realization that the group of motions allows very few changes in properties leads to the need to investigate more general types of transformations that do not leave as many invariant properties. One such group, the similarities, is introduced in the last section of this chapter, following the section on motions in three-space.

EXERCISES 2.5

1. Draw a figure for each case in Table 2.2.

2. Verify, as suggested, that the equations for direct motions have a $+$ symbol and the equations for opposite motions have a $-$ symbol.

3. Draw a specific example showing that the product of a reflection and a rotation could be a glide reflection.

4. Draw a specific example showing that the product of a reflection and a rotation could be a reflection.

5. Draw a specific example showing that the product of two glide reflections could be a translation.

6. Show analytically that a translation has no invariant points.

7. Investigate analytically the results of setting $x = x'$ and $y = y'$ in the equations of a rotation.

8. Use Theorem 2.6 to prove that, if three noncollinear points of a plane remain invariant under a motion, the motion is the identity.

9. Prove analytically that, for a translation, the image of two intersecting lines is two intersecting lines and the image of the original point of intersection is the new point of intersection.

10. Prove analytically that the angle between two intersecting lines is an invariant under the set of all translations.

2.6   MOTIONS OF THREE-SPACE

The motions of three-space are less familiar than those of two-space. To discover the various types of motions, it is again essential to see what types of transformations can be applied to a set of points in three-space so that the distance between any two points is preserved. These transformations are introduced rather briefly in this section.

The motions of three-space are six in number. The three simplest motions are translation, rotation, and reflection, but these concepts must be clarified for three-space.

The definition of translation in space is analogous to the definition of translation for the plane.

DEFINITION. The transformation T_A such that $T_A(X) = X + A$ is a translation of the space in the direction OA.

For example, in Figure 2.30, if A is the point (2, 4, 8) and X is the point (1, 2, 3), then $T_A(X) = (3, 6, 11)$. The image point can easily be found by addition of coordinates. Verify the fact that a translation in three-space sends all the points of a plane into the points of a parallel plane.

FIGURE 2.30

DEFINITION. The rotation $R(l, \alpha)$ in three-space through an angle α about a fixed line is a rotation in a plane perpendicular to l. See Figure 2.31.

FIGURE 2.31

The image of point A not on l is a point B such that $\overleftrightarrow{AO}$ and $\overleftrightarrow{OB}$ determine a plane π perpendicular to l.

DEFINITION. A reflection R_π in three-space is a reflection about a plane such that, for P and its image P', π is the perpendicular bisector of $\overline{PP'}$. See Figure 2.32.

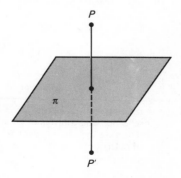

FIGURE 2.32

The complete analytic presentation of motions for three dimensions is not given, but special cases are considered. The following equations for a translation in three-space are analogous to those for translations in a plane:

$$x' = x + a,$$
$$y' = y + b,$$
$$z' = z + c.$$

For a rotation in space, if the fixed line is the z-axis and the plane is the xy-plane, then the image of a point in that plane can be found using the rotation equations given previously for rotation in a plane about the origin. For a point in a plane parallel to the xy-plane, the z coordinate will remain constant.

For a reflection about the xy-plane, the image of (a, b, c) is $(a, b, -c)$, hence

$$x' = x,$$
$$y' = y,$$
$$z' = -z.$$

Corresponding sets of equations can be written for reflections about the xz- and yz-planes.

In addition to the three simple motions, it is necessary to define three additional types of motions which are products of the basic ones so that the set of motions of three-space will be closed. Table 2.6 shows the definitions of the additional motions as products of two simpler motions.

TABLE 2.6

Motion	Explanation
a. screw displacement	rotation and translation along axis of rotation
b. glide reflection	reflection and translation parallel to plane of reflection
c. rotatory reflection	reflection and rotation with axis perpendicular to plane of reflection

A complete analysis of the motions of three-space is not given here, but the theorems are analogous to the theorems in two dimensions.

THEOREM 2.7. The motions of three-space constitute a group of transformations.

While the complete proof is not given here, it is analogous to the proof of Theorem 2.4 in that the analysis depends on classifying motions in three-space as direct or opposite and showing that each space motion can be expressed as the product of reflections about a plane. Table 2.7 shows which motions are direct and which are opposite. It also gives a description of the invariant points for each.

TABLE 2.7

Motion in Three-Space	Direct or Opposite	Invariant Points
rotation	direct	a line
translation	direct	none
reflection	opposite	plane of reflection
rotatory reflection	opposite	at least one point
screw displacement	direct	none
glide reflection	opposite	none

The table makes it possible to indicate the product of any two motions, but the results cannot be determined uniquely without more information. For example, the product of a screw displacement and a glide reflection is an opposite transformation, hence it is either a reflection, a rotatory reflection, or a glide reflection.

The study of Table 2.7 shows that a reflection has an invariant plane and that no other motion has as extensive a set of invariant points. The following theorem is analogous to Theorem 2.6.

THEOREM 2.8. A motion in space is uniquely determined by a tetrahedron and its image.

The proof is left as an exercise.

EXERCISES   2.6

1. Let the vector for a translation be $(-2, 1, 7)$, and give the image for each point under the translation.

a. $(0, 0, 0)$ b. $(2, 5, 3)$
c. $(-2, -3, -5)$ d. $(0, -4, 1)$

2. In the translation with vector $(-1, -3, 4)$, what points have these images?

    a. $(0, 0, 0)$ b. $(2, -1, 3)$

3. Describe the inverse for each type of motion of space.

4. What is the image of the point $(2, 4, 5)$ under a rotation of $45°$ in the plane $y = 4$ about the y-axis?

For Exercises 5–8, give the image of the point under a reflection about the given plane.

5. $(3, 8, 1)$, xy plane   6. $(-2, 4, 3)$, xz plane
7. $(2, 3, 9)$, yz plane   8. $(-1, 4, -2)$, $x = 3$

9. Sketch an example of each of the last three types of motions of space.

10. Prove that, if four noncoplanar points of space remain invariant under a motion, the motion is the identity.

11. Name the possibilities for the type of motion that is the product of:

    a. A rotatory reflection and a translation.
    b. A rotation and a glide reflection.

12. Derive the equations for a reflection about the plane $z = 2$.

13. Prove Theorem 2.8.

Make a sketch showing a tetrahedron and its image for a:

14. Translation.

15. Rotation.

16. Reflection.

2.7   SIMILARITY TRANSFORMATIONS

In the Euclidean motions studied in the last three sections of this chapter, distance was preserved. Properties such as length, area, and volume are all preserved under isometries. Even in high school

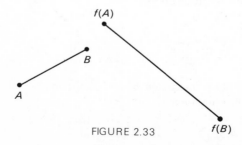

FIGURE 2.33

geometry, however, the idea of similar but noncongruent figures is introduced. Similarity is an example of a type of transformation in which distance between points is not preserved. But distance between points is modified in a consistent way throughout the plane.

DEFINITION. A plane similarity is a transformation of the plane onto itself such that $|f(A)f(B)| = r|AB|$, where r is some positive real number. See Figure 2.33.

For this example, $r = 2$, so that $\overline{AB}$ is stretched by the similarity transformation into a segment twice its length. For $r = 1/3$, it would have been shrunk into a segment one third of its length. For $r = 1$, the similarity transformation is an isometry. The number r is called the *ratio of similarity*.

THEOREM 2.9. The set of all similarities of the plane is a transformation group.

The inverse of a similarity with ratio r has the ratio $1/r$. The ratio of a product of two similarities is the product of the ratios of the two similarities. For example, if the first ratio of similarity is two and the second ratio of similarity is three, then the ratio of similarity for the product transformation is six. Theorem 2.9 makes it possible to think of the motions of a plane as a subgroup with $r = 1$ of the group of similarities of the plane.

In the previous section, it was proved that a unique motion is determined if three noncollinear points and their images are known. The same proof can be modified to apply to the more general group of similarities.

THEOREM 2.10. A plane similarity is uniquely determined when a triangle and its image are given.

In Figure 2.34, $\triangle A'B'C'$ is the image of $\triangle ABC$. The distances $P'A'$, $P'B'$, and $P'C'$ are determined, and P' is a uniquely determined point for any given point P.

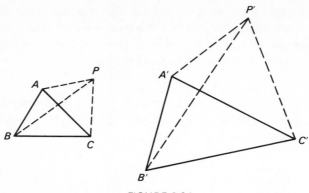

FIGURE 2.34

An intuitive exploration of how to explain similarities as a series of simple steps leading from a figure to its image indicates that there must be both a change in position and a uniform change in size (but not a change in shape).

For example, in Figure 2.34, changes of position can result in triangle ABC being situated so that A and A' coincide, so that $\overline{AB}$ lies along $\overline{A'B'}$, and so that $\overline{AC}$ lies along $\overline{A'C'}$. A change in size is then necessary if B and B' are to coincide. Since motions of the plane can accomplish the change in position, the new transformation needed is one to accomplish the uniform change in size. The new type of transformation must be a special similarity in which the image of a set of points is a similar figure with corresponding sides parallel.

DEFINITION. A *homothety* $H(O,r)$ is a similarity such that, for any point P, $OP' = rOP$ where O is the center of the homothety and r is its ratio.

The center of homothety is an invariant point, according to this definition. The ratio of the distances from the center to the image point and to the original point is a constant, the ratio for the homothety. For example, Figure 2.35 shows a homothety with $OP'/OP = 3/2$. Remember that the ratio of similarity always compares the image with the original.

The exploration of the need for a homothety showed that motions could result in an image in the correct relative position; thus cor-

FIGURE 2.35

responding segments under a homothety should be parallel. That the property of parallelism is an invariant under a homothetic transformation is proved as the next theorem.

THEOREM 2.11. The image of a segment under a homothety is a parallel segment.

In Figure 2.35, $OP'/OQ' = OP/OQ$, and $\triangle OP'Q' \sim \triangle OPQ$, hence $\overline{PQ} \parallel \overline{P'Q'}$.

It is sometimes said that homothetic figures are both similar and similarly placed, since their corresponding sides are parallel. For example, in Figure 2.36, $\triangle ABC$ and $\triangle A'B'C'$ are similar but not homothetic, whereas $\triangle ABC$ and $\triangle A''B''C''$ are homothetic. Also, $\triangle A'B'C'$ and $\triangle A''B''C''$ are similar but not homothetic.

FIGURE 2.36

The property of parallelism is preserved under a homothety. Another invariant property is the property of being a circle.

THEOREM 2.12. The image of a circle under a homothety is a circle. See Figure 2.37.

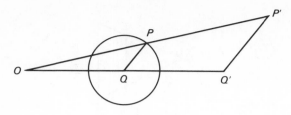

FIGURE 2.37

Let O be the center of homothety and P' and Q' the images of a point on the given circle and its center, respectively. Since $\overline{PQ} \parallel \overline{P'Q'}$ by Theorem 2.11, $P'Q'/PQ = OQ'/OQ$, or

$$P'Q' = \frac{PQ \cdot OQ'}{OQ}.$$

But each segment in the expression on the right has a fixed length, hence $P'Q'$ is a constant. For any position of point P, P' lies on a circle with Q' as center.

So far it has been determined that a similarity can be considered the product of a homothety and a motion. Because a homothety also can accomplish a translation, the following stronger statement can be established.

THEOREM 2.13. A plane similarity that is not a motion is the product of a homothety and a rotation or a reflection.

It is assumed here, and can be proved using theorems from topology, that every plane similarity not a motion has exactly one fixed point. In Theorem 2.13, the rotation is about the fixed point and the reflection is about a line through the fixed point.

In Figure 2.38, let O be the fixed point and $\overline{AB}$ any segment. A' and B' are the images of A and B under a homothety with O as center. Let $f(A)$ and $f(B)$ be the images of A and B under any similarity that is not a motion. $\triangle OA'B' \sim \triangle Of(A)f(B)$. $\triangle OA'B'$ can be mapped into $Of(A)f(B)$ either by a rotation through angle $AOf(A)$ or by a reflection about the bisector of this angle. In the first case, the similarity is direct, and in the second case the similarity is opposite.

The analytic development of the equations for similarity trans-

FIGURE 2.38

formations begins with the observation that the equations for a homothety with center at the origin are

$$x' = rx,$$

$$y' = ry.$$

Since a similarity is the product of a motion and a homothety, the equations for a similarity are of the form

$$x' = k(ax + by + c)$$

$$y' = \pm k[(-bx + ay) + d]$$

for $a, b, c, d \in R$ and $a^2 + b^2 = 1$.

Similarities in three space also form a group having the motions of three-space as a subgroup. The theory and definitions are analogous to those for two dimensions.

THEOREM 2.14. A space similarity is uniquely determined when a tetrahedron and its image are given.

THEOREM 2.15. The set of all similarities of space is a transformation group.

THEOREM 2.16. A space similarity that is not a motion of space is the product of a homothety about its fixed point and a rotation about a line passing through the fixed point.

For Theorem 2.16 to be true, it is necessary to reinterpret a homothety in a somewhat more general way by allowing the ratio r to be either positive or negative. The physical interpretation of a negative ratio is that a point and its image are on opposite sides of the center of homothety.

Consider, because of Theorem 2.14, any tetrahedron $OABC$ with the fixed point O as one vertex. It is assumed here, as for two dimensions, that a fixed point exists for any similarity. Let r be the ratio of similarity. Then the image of $OABC$ under the homothety $H(O,r)$, $OA''B''C''$, is congruent to the image $OA'B'C'$ under the similarity. If the motion connecting $OA''B''C''$ and $OA'B'C'$ is direct, then the similarity is the product of $H(O,r)$, for r positive, and a rotation about some line through O. If the similarity is opposite, then r is negative.

This chapter has enlarged the idea of Euclidean geometry by showing how it can be considered the study of those properties that are invariant under the group of motions (or under the group of similarities). Through this new approach, additional significance is attached to the study of congruent and similar triangles. Even more important, however, is the anticipation that more general transformations exist (such as those given in Chapter 7, for example), with Euclidean geometry as a special case.

The study of Euclidean geometry has been revitalized during the present century. In addition to the approach through transformations, a classification and detailed study of sets of points in Euclidean geometry utilizing the modern concept of convexity is a subject of current research. The geometry of convexity is the subject of Chapter 3.

EXERCISES   2.7

1. Find the length of the image of a 3-inch segment under a similarity with ratio 4/3.

2. Find the ratio of similarity if a 5-inch segment has a 6-inch image.

3. Name some properties preserved under all motions which are not preserved under all similarities.

4. What is the effect of a similarity on the area of a triangular region?

5. Find the ratio of similarity if a square region with an area of 12 square units has an image with an area of 17 square units.

6. In Figure 2.37, compare the ratio of the radii of the two circles with the ratio of similarity.

7. Find the image of $(3, 5)$ under a homothety with center at $(0, 0)$ and with ratio 3/4.

8. Find the image of $(4, 3)$ under the similarity with equations

$$x' = 2(3x - y + 5),$$
$$y' = 2(x + 3y + 2).$$

9. Prove analytically that if two lines intersect, their images under a similarity also intersect.

10. Is the set of homotheties with a given invariant point a group of transformations?

11. Prove analytically that the property of being a parabola is an invariant for the group of homotheties with a given invariant point.

CHAPTER 3

CONVEXITY

3.1 BASIC CONCEPTS

The study of convexity in Euclidean geometry is a very modern development. As is true with the approach through transformations, the study of convex sets employs many basic ideas of modern mathematics to provide added meaning to concepts that have long been a part of geometry.

The words "convex" and "convexity" are common outside of mathematics. *Convex* means curving outward. For example, a convex lens bulges outward. Fortunately, the use of the word *convex* to describe particular sets of points in mathematics is only a careful refinement of this common meaning. Study Figure 3.1, parts a and b, to discover the basic difference between convex sets of points and those that are nonconvex.

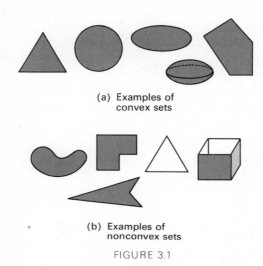

(a) Examples of
convex sets

(b) Examples of
nonconvex sets

FIGURE 3.1

For a convex set of points, such as K in Figure 3.2a, any segment with endpoints A and B in the set lies wholly in the set. This is not true for all pairs of points in the nonconvex set in Figure 3.2b. For example, $\overline{CD}$ includes points outside set K'.

(a)   (b)

FIGURE 3.2

DEFINITION. A *convex set* is a set of points K such that if $A, B \in K$, then $\overline{AB} \subset K$.

The symbol $\subset$ means "is a subset of." Although this definition of convex set may seem restrictive, it is general enough to include many of the common sets of points studied in Euclidean geometry. Examples are a circular region, some polygonal regions, segments, an angle (with measure less than π) and its interior, and a spherical region. The empty set and a set consisting of a single point are both convex by agreement.

To show that a particular set is convex by definition is not always easy. In addition to Theorem 3.1 below, analytic geometry is often employed. You should assume here and elsewhere in this chapter, unless stated to the contrary, that the variables for coordinates of points are elements of the set of real numbers.

The postulational system for a secondary school geometry includes the assumption that a half-plane is a convex set. For example, see Appendix 3, Postulate 14. The determination of whether a set is convex or not is facilitated by the following theorem.

THEOREM 3.1. The intersection of two convex sets is a convex set.

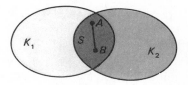

FIGURE 3.3

In Figure 3.3, let K_1 and K_2 be any two convex sets and let set S be their intersection. For A, $B \in S$, all points of $\overline{AB}$ are elements of K_1 because K_1 is convex. All points of $\overline{AB}$ are also elements of K_2, because it is also convex. Hence, $AB \subset (K_1 \cap K_2)$; thus, $K_1 \cap K_2$ is convex. Theorem 3.1 may be generalized to more than two sets, as in Exercise 3 of Exercise Set 3.1. The theorem may also be used to show the convexity of other common sets of points. For example, an angle (measure less than π) and its interior can be defined as the intersection of two half-planes, so that Theorem 3.1 applies.

An alternative approach is the use of analytic geometry to prove that sets of points are convex. This approach can even be used for a half-plane.

EXAMPLE. Show that $S = \{(x, y): x > 0\}$ is a plane convex set. (S is the set of points with coordinates (x, y) such that $x > 0$.)

Let $A(x_1, y_1)$ and $B(x_2, y_2)$ be any two points of S, as in

Figure 3.4, such that $x_1 > 0$, $x_2 > 0$, and $x_2 \geqq x_1$. Then, for any point $P(x_k, y)$ of $\overline{AB}$, $x_2 \geqq x_k \geqq x_1$, hence $x_k > 0$ and $\overline{AB} \subset S$, so that S is convex by definition.

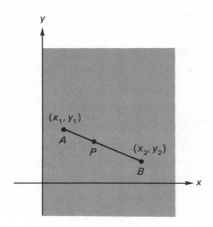

FIGURE 3.4

Credit for beginning the systematic study of convex sets in the early twentieth century is given to H. Brunn and H. Minkowski. A 1934 survey by T. Bonnesen and W. Fenchel showed much progress. The Proceedings of Symposia in Pure Mathematics, Volume VII, was the publication resulting from the 1963 Symposium on Convexity sponsored by the American Mathematical Society. This volume is an interesting reference text for the student of convexity.

Before exploring more complex properties of convex sets, it is necessary to introduce several basic concepts from modern geometry that will be useful both here and in later chapters.

Recall that a function $y = f(x)$ is continuous at $x = a$ if and only if, given $e > 0$, there exists a δ such that

$$|f(x) - f(a)| < e \quad \text{if} \quad |x - a| < \delta.$$

A function is continuous if it is continuous at every point of its domain.

For example, the function in Figure 3.5a is continuous, whereas the function in Figure 3.5b is not.

A *curve* is the graph of a set of equations of the form $x = f(t)$, $y = g(t)$, for f and g continuous functions and the domain of t

(a)   (b)

FIGURE 3.5

an interval of real numbers. If the function is one-to-one, except possibly at the endpoints, the curve is a•*simple curve,* and if the points corresponding to both endpoints of the defining interval are identical, the curve is a *closed curve.* For this section, you will need to recognize examples of these various types of curves from drawings so that the equation need not be dealt with. Intuitively, a simple closed curve is thought of as a curve that begins and ends at the same point but does not cross itself; thus there is only one interior. (See Figure 3.6.) The set of points on and inside a simple closed curve or an angle in a plane is called a *plane region.*

Curve   Simple curve   Closed curve   Simple closed curve

FIGURE 3.6

Precise definitions of interior, exterior, and boundary points of a set of points such as a plane region depend on the concept of neighborhood of a point.

DEFINITIONS

For Two Dimensions: The *open circular neighborhood* with radius r of a point P is the set of points inside a circle of radius r with P as center. (See Figure 3.7a.) This definition can be written in symbols as

$$N(P,r) = \{A: |PA| < r\}.$$

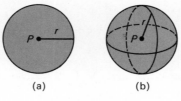

(a)   (b)

FIGURE 3.7

For Three Dimensions: The *open spherical neighborhood* with radius r of a point P is the set of points in space inside a sphere of radius r with P as center. (See Figure 3.7b.) This definition can be written in symbols as

$$N(P,r) = \{A: |PA| < r\}.$$

A *closed neighborhood* in two or three dimensions includes the points on the circle or sphere as well as those inside. The definition of closed neighborhood, using set symbolism, is

$$N[P,r] = \{A: |PA| \leqq r\}.$$

Both open and closed neighborhoods are convex sets of points.

DEFINITION. P is an *interior point* of a set S if and only if there is a positive real number r such that $N(P,r) \subset S$.

This definition means intuitively that a point is an interior point of a set if every point sufficiently near it is an element of the set. For example, point A is an interior point in Figure 3.8a. This definition

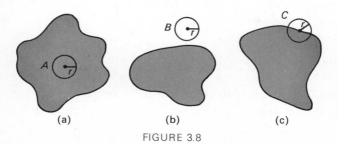

(a)   (b)   (c)

FIGURE 3.8

of interior agrees with the common understanding of interior that is used in speaking of the interior of a simple closed curve or the interior of a sphere.

DEFINITION. P is an exterior point of a set S if and only if there is a positive real number r such that $N(P,r) \subset \sim S$.

The symbol $\sim S$ means "not S" or the complement of S. A point is an exterior point of a set if every point sufficiently near it is a member of the complement of the set. For the set of points $x^2 + y^2 < 1$, the exterior is $x^2 + y^2 > 1$. Point B is an exterior point in Figure 3.8b.

A point that is neither an interior nor an exterior point of a set is called a *boundary point*. The set of boundary points for a set is called the *boundary* of the set. Note that if P is a boundary point of S, then it is also a boundary point of the complement of S.

DEFINITION. P is a boundary point of set S if *every* neighborhood of P contains both points of S and of $\sim S$.

For the set $x^2 + y^2 < 1$, the boundary is $x^2 + y^2 = 1$. Point C is a boundary point in Figure 3.8c. The boundary of a half-plane is the line determining the half-plane. A simple closed curve is the boundary of the set of points in its interior.

You should realize that the boundary points of a set may or may not be elements of the set. Another important conclusion is that,

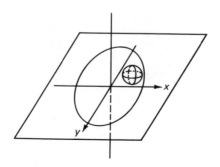

FIGURE 3.9

to decide whether a point is an interior, an exterior, or a boundary point, it is necessary to consider the number of dimensions. For example, in Figure 3.9, $\{(x, y, 0): x^2 + y^2 \leq 1\}$ has no interior points in three-space. Each point of the set is a boundary point, since every spherical neighborhood for a point of the set contains both points of the set and points of its complement.

The classification of points as interior, exterior, or boundary points for a set leads to useful classifications of the sets themselves.

DEFINITION. An *open* set has only interior points.

An open neighborhood is an open set, as is the interior of any simple closed curve.

DEFINITION. A *closed* set contains all its boundary points.

Examples of closed sets are $x^2 + y^2 \leq 1$, polygonal regions, and a segment $\overline{AB}$. It is important to observe that these definitions of open and closed sets are not mutually exclusive, nor do they include all possible sets of points. A set may be both open and closed or neither open nor closed. An example of a set both open and closed is the entire plane. Since this set has no boundary, it includes all its boundary points. At the same time, all of its points are interior points. Examples of sets that are neither open nor closed are given in Figure 3.10.

FIGURE 3.10

DEFINITION. A *bounded set* in two dimensions is one that is a subset of some circular region with a real number for radius. In three dimensions, a bounded set is a subset of some spherical region with real radius.

Some bounded sets are shown in Figure 3.11. Bounded sets cannot extend indefinitely. For example, the parabola $y = x^2$ is not a

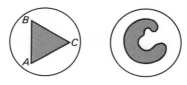

FIGURE 3.11

bounded set because no circle can enclose the set of points. The concept of a bounded set should not be confused with the boundary of a set.

EXERCISES   3.1

1. Which of these sets are convex?

   a. An angle.
   c. Interior of an ellipsoid.
   e. A single point.
   g. A triangular region.

   b. A rectangular region.
   d. Straight line.
   f. A ray.
   h. A triangle.

2. Which of these sets are convex sets?

   a. A circular region with one point on the boundary removed.
   b. A rectangular region with one vertex removed.
   c. A rectangular region with one point, not a vertex, on the boundary removed.
   d. A circular region with one interior point removed.

3. Prove that the intersection of any collection of convex sets is a convex set.

4. Is the union of two convex sets ever a convex set? Is it always a convex set?

5. Prove that a triangular region is a convex set.

6. Show analytically that $S = \{(x, y): x > 3\}$ is a convex set.

7. Show analytically that $T = \{(x, y): x > 3 \text{ and } y > 4\}$ is a convex set.

8. Give a definition for a one-dimensional neighborhood.

9. Describe the interior, boundary, and exterior for these plane sets.

   a. $\{(x, y): x^2 + y^2 \leq 1\}$
   c. $\{(x, y): y < |x|\}$

   b. $\{(x, y): y > x^2\}$
   d. $\left\{(x, y): x^2 + \dfrac{y^2}{4} \leq 1\right\}$

e. $\{(x, y): x$ and y are integers$\}$
f. $\{(x, y): x$ and y are rational numbers$\}$

10.   Describe the interior, boundary, and exterior for these sets of points in space.

a. $\{(x, y, z): x^2 + y^2 + z^2 \leq 1\}$ b. $\{(x, y, 0): x^2 + y^2 \leq 9\}$
c. $\{(x, y, 0): x > y\}$ d. $\{(x, y, 0): x = 2\}$

11.   Classify these sets as open, closed, neither, or both.

a.   The sets in Exercise 9.   b.   The sets in Exercise 10.
c.   A rectangular region with one vertex removed.

12.   Which of the sets in each exercise are bounded?

a.   The sets in Exercise 9.   b.   The sets in Exercise 10.

13.   Prove that the boundary of a set of points is also the boundary of the complement of the set.

14.   Prove that the complement of a closed set is open.

15.   What can you say about the complement of a set that is neither closed nor open?

16.   Give examples of (a) a bounded set that does not contain its boundary and (b) a set that contains its boundary but is not bounded.

3.2   CONVEX SETS AND SUPPORTING LINES

The basic ideas of convex sets were introduced in the previous section, along with the vocabulary needed to continue the study. It is now possible to explore some additional concepts associated with convex sets. In this section, it is assumed that the geometry being considered is the geometry of two-space. A concept especially important in discussing convex sets, but applicable to other sets of points also, is the concept of supporting line.

DEFINITION. A *supporting line* for a set in two dimensions with interior points is a line through at least one of the boundary points of the set such that all points of the set are in the same closed half-plane determined by the line.

Figure 3.12 shows three examples of sets, with several supporting lines for each.

FIGURE 3.12

The proof of the following theorem results in an alternative definition for supporting lines of convex sets in two dimensions with interior points.

THEOREM 3.2. A line is a supporting line for a convex set of points if it goes through at least one boundary point of the set but no interior points, and conversely.

Theorem 3.2 includes two implications:
a. If a line goes through at least one boundary point of the set but contains no interior points, then it is a supporting line for the convex set of points.
b. If a line is a supporting line for a convex set of points, then it contains at least one boundary point of the set but no interior points.

Associated with each implication are three other implications, the converse, inverse, and contrapositive. These are shown symbolically as follows:

Original implication	$p \rightarrow q$
Converse	$q \rightarrow p$
Inverse	$\sim p \rightarrow \sim q$ (not p implies not q)
Contrapositive	$\sim q \rightarrow \sim p$

An implication and its contrapositive are logically equivalent. So are the converse and the inverse. Furthermore, the converse of one implication of Theorem 3.2 is the other implication. Proving an "if and only if" theorem is the same as proving a theorem and its converse. Proving $p \rightarrow q$ and $q \rightarrow p$ results in proving the equivalence $p \leftrightarrow q$.

Sometimes, especially in geometry, it is easier to prove the inverse and/or contrapositive than some other form of the implication.

Any of these pairs of statements may be proved to prove an implication and its converse:

$$\begin{pmatrix} \text{Theorem} \\ \text{Converse} \end{pmatrix} \begin{pmatrix} \text{Theorem} \\ \text{Inverse} \end{pmatrix} \begin{pmatrix} \text{Contrapositive} \\ \text{Inverse} \end{pmatrix} \begin{pmatrix} \text{Contrapositive} \\ \text{Converse} \end{pmatrix}$$

For the first implication of Theorem 3.2, the easiest to prove of these pairs is that including the contrapositive and the inverse.

Contrapositive	*Inverse*
If a line is not a supporting line for a convex set in two dimensions but contains a boundary point, then it contains interior points	If a line contains at least one boundary point and also interior points, then it is not a supporting line for the convex set in two dimensions.

FIGURE 3.13

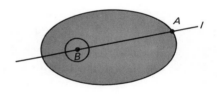

FIGURE 3.14

Proof: If *l* is not a supporting line, as it is not in Figure 3.13, then points *A* and *B* of set *K* can be found in different half-planes formed by *l*. The intersection of $\overline{AB}$ and *l* is not empty. But *C* is an interior point of set *K*, since *K* is convex and *A* and *B* are interior points. See Exercise 14, Exercise Set 3.2.

Proof: Suppose *l* contains boundary point *A* and interior point *B*, as in Figure 3.14. There exists a neighborhood of *B*, $N(B, e)$, containing interior points. But this neighborhood includes points in both of the half-planes formed by *l*; hence, *l* is not a supporting line.

The concept of supporting line is closely related to the more familiar idea of a *tangent*. From calculus, the intuitive idea of a tangent to a curve is that of a line intersecting the curve at a point and having the same slope as the curve at that point. As Figure 3.15 shows, supporting lines and tangents are not necessarily the same thing.

(a) A tangent that is not   (b) A supporting line
    a supporting line   that is not a tangent

FIGURE 3.15

In convex geometry, the concept of tangent, which does not depend on the notion of limits, is somewhat different from the concept used in calculus. The beginning concept is that of a tangent cone.

DEFINITION. A *tangent cone* of a convex set is the set of all rays that: (a) have boundary points of the set as endpoints and (b) also pass through other points of the convex set or its boundary.

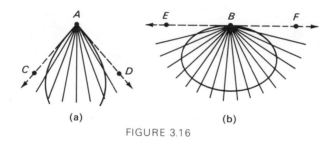

(a)   (b)

FIGURE 3.16

Figure 3.16 shows two tangent cones. For Figure 3.16a, $\overrightarrow{AC}$ and $\overrightarrow{AD}$ are the boundaries of the tangent cone. (The notation $\overrightarrow{AC}$ means ray AC.) In general, the boundaries are not themselves rays of the tangent cone. The boundaries of the tangent cone are called *semitangents*. In Figure 3.16b, $\overrightarrow{BE}$ and $\overrightarrow{BF}$ are semitangents and are collinear.

DEFINITION. The union of two collinear semitangents is the *tangent* to a convex set at a point.

From Figure 3.16, or from a consideration of the definition of tangent, it should be clear that there are tangents at some boundary points of a convex set and not at others. It is possible to classify boundary points of a convex set as *regular* or *corner* points on the basis of whether there is or is not a tangent to the curve passing through that boundary point. There is a tangent at every regular point. In Figure 3.16, A is a corner point, whereas B is a regular point. Other examples of corner points are the vertices of a convex polygonal region. On the other hand, all of the points on a circle are regular points on the boundary of the circular region.

One of the main reasons for introducing the concept of supporting line is that it can be used to distinguish a convex set from one that is nonconvex.

THEOREM 3.3. The interior of a simple closed curve is a convex set if and only if through each point of the curve there passes at least one supporting line for the interior.

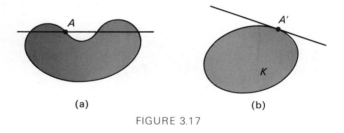

(a)   (b)

FIGURE 3.17

Figure 3.17 helps to explain intuitively the significance of Theorem 3.3. If a set is not convex, then points (such as A) exist on the boundary such that any line through A contains interior points. Through any point A' on the boundary of a convex set, at least one line can be drawn containing no interior points.

Proof:

a. If K is convex, then there is at least one supporting line passing through each boundary point A.

Let $\overrightarrow{AB}$ and $\overrightarrow{AC}$ be the semitangents at point A in Figure 3.18. If the measurement of $\angle BAC$ is π, then $\overleftrightarrow{BC}$ is a tangent and hence a supporting line. If the measurement of $\angle BAC$ is less than π, then $\overleftrightarrow{AB}$ and $\overleftrightarrow{AC}$ are both supporting lines. If the measurement of $\angle BAC$

is greater than π, then the original set is not convex, contrary to the beginning assumption. See Exercise 12, Exercise Set 3.2.

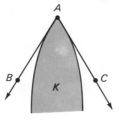

FIGURE 3.18

b. If S' is not a convex set, then there is not a supporting line at every boundary point.

This statement is the inverse of the implication in (a), so proving it will complete the proof of Theorem 3.3. (See Figure 3.19.) Since S' is not convex, an interior point C and a boundary point B can be found such that a boundary point A lies on $\overline{BC}$. $\overleftrightarrow{BC}$ is not a supporting line, since it contains interior points. Any other line through A has B and C in opposite half-planes; hence it cannot be a supporting line. There is no supporting line to S' through A.

FIGURE 3.19

If a line is a supporting line for a set in two dimensions, the closed half-plane formed by the line and containing the set is called a *supporting half-plane*. This concept can also be used to give an additional property of convex sets.

THEOREM 3.4. A plane closed convex set that is a proper subset of a plane is the intersection of all its supporting half-planes.

An example of this theorem for a quadrilateral and its interior, showing four of the supporting planes formed by the sides, is in Figure 3.20.

FIGURE 3.20

Theorem 3.4 states that the set of points in the convex set is identical to the set of points in the intersection of the supporting half-planes. A convenient way to prove that two sets are identical is to prove that each is a subset of the other, since the sets must then have exactly the same elements. Let the convex set be K and the intersection of the supporting half-planes be K'. By definition of supporting half-planes, K is a subset of K'.

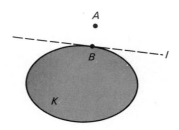

FIGURE 3.21

Now let A be a point in the complement of K, as in Figure 3.21. It can be proved that some point of K, say B, is the closest point of K to A. Also, the supporting line l to K at B is perpendicular to $\overleftrightarrow{AB}$. Then A is not in the supporting half-plane of K determined by l, therefore $A \notin K'$. Since any point in the complement of K is also in the complement of K', $K' \subseteq K$, hence $K = K'$.

EXERCISES   3.2

For Exercises 1–4, tell whether the line is or is not a supporting line for the set.

Set	Line
1. Circular region	line through center
2. Circular region	tangent
3. Square region	line containing side of square
4. Square region	line containing diagonal of square

For Exercises 5–8, describe the regular points and corner points for each of these sets.

5. Circular region.

6. Square region.

7. Angle and its interior—measure less than 180°.

8. Convex polygonal region.

9. Use Theorem 3.4 to give a definition of:

   a. A triangular region.   b. A convex polygonal region.

10. Draw three nonconvex sets and show one point on the boundary of each through which no supporting line passes.

11. State the converse, inverse, and contrapositive of the second implication in Theorem 3.2.

12. Prove that the interior angles of a convex polygon have measurements less than π.

13. Prove that a tangent cone to a convex set at a boundary point is itself a convex set.

14. Prove that if A is an interior point of convex set K and B is any other point of K, then every point in $\overline{AB}$ ($\overline{AB}$ except for point B) is an interior point of K. Hint: Use Theorem 3.3.

15. Prove that if A and B are boundary points of a convex set K as well as points of K, then either $\overline{AB}$ ($\overline{AB}$ without the endpoints) is entirely in the interior of K or $\overline{AB}$ is entirely in the boundary of K.

3.3   CONVEX BODIES IN TWO-SPACE

In the previous section, the concept of supporting line gave added meaning to convexity. Many of the common sets of points from Euclidean geometry were found to be convex sets. These ideas can now be used to define a special type of convex set called a convex body.

DEFINITION. A *convex body* is a convex set of points that is closed, bounded, and nonempty.

Many common convex sets are convex bodies. Examples include convex polygonal regions and line segments. On the other hand, the interior of a parabola and the entire plane are convex sets that are not convex bodies. Convex bodies always have interior points for the number of dimensions considered.

The following theorem gives an important characteristic of a two-dimensional convex body.

THEOREM 3.5. A simple closed curve S and its interior form a convex body K if and only if every line through an interior point of K intersects S in exactly two points.

Proof:

a. If a simple closed curve S is the boundary of a convex body K, then every line through an interior point of K intersects S in exactly two points.

See Figure 3.22. Let A be an interior point of K. Any ray $\overrightarrow{AP}$ intersects K in a segment $\overline{AB}$. This is true because the intersection of two convex sets is a convex set. $\overline{AB}$ is a one-dimensional convex body. But $\overrightarrow{AB'}$, where B', A, and B are collinear, also intersects K in a segment $\overline{AB'}$, and B' is the second of the two boundary points on the line through A.

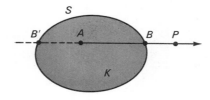

FIGURE 3.22

b. If every line through each interior point of K intersects the simple closed curve S in exactly two points, then S is the boundary of a convex body K. (It can be shown that the contrapositive of this second implication, which is logically equivalent to the implication itself, holds.

The contrapositive states that if K is nonconvex, then some line through an interior point of K does not intersect the boundary S in exactly two points.)

In Figure 3.23, if K is nonconvex, interior points A and B can be found such that C on $\overline{AB}$ is an exterior point. Then it can be assumed that boundary points D and E for K exist on $\overline{AB}$. The ray $\overrightarrow{DA}$ also has a third boundary point F, so $\overleftrightarrow{AB}$ intersects S in at least three points.

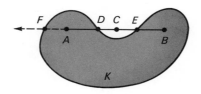

FIGURE 3.23

Another theorem about a convex body and its boundary is assumed here, although a topological proof has been provided by Verner Hoggatt, Jr.

THEOREM 3.6. The boundary of a convex body in two-space is a simple closed curve.

Of course, the converse of Theorem 3.6 is not true, because simple closed curves include a great variety of shapes in addition to those that are convex. Another condition for determining whether a particular simple closed curve is the boundary of a convex body is provided by the next theorem.

THEOREM 3.7. A simple closed curve S is the boundary of a two-dimensional convex body K if and only if each closed polygon $T = P_0 P_1 \ldots P_n P_0$ determined by successive points on and inscribed in K is the boundary of a convex polygonal region.

The significance of the theorem is illustrated in Figure 3.24. Any polygonal region inscribed in the simple closed curve of Figure 3.24a will be convex. For the closed curve in Figure 3.24b, however, some of the inscribed polygonal regions, such as the one shown, are nonconvex.

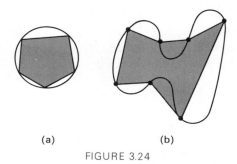

(a)   (b)

FIGURE 3.24

The proof of the "only if" implication involves showing that any inscribed polygon for a convex body has a supporting line at each point.

Let l be the line containing one side $\overline{AB}$ of an inscribed polygon. Since K is convex, $\overline{AB} \subset K$, so that $\overset{\circ}{\overline{AB}}\!\!\overset{\circ}{}$ contains only points of S or only interior points of K (Exercise 15, Exercise Set 3.2). If $\overset{\circ}{\overline{AB}}\!\!\overset{\circ}{}$ contains only points of S, then it can contain no interior points of T and is a supporting line for both K and T. If $\overset{\circ}{\overline{AB}}\!\!\overset{\circ}{}$ contains interior points of K, as in Figure 3.25, then S is partitioned into two curves lying on opposite sides of l. All the vertices of T lie in one supporting half-plane H determined by l, since otherwise A and B would not be successive vertices. Then l is a supporting line for $H \cap K$ and thus is a supporting line for T and its interior. Every point of T has a supporting line passing through it, so that T is the boundary of a convex polygonal region.

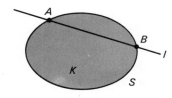

FIGURE 3.25

The proof of the if statement can be accomplished by showing that, if the set K with boundary S is not convex, there exists an inscribed quadrilateral that is not the boundary of a convex region. Such a quadrilateral is pictured as $ABCD$ in Figure 3.26, but the details of the proof are left as Exercise 10 of Exercise Set 3.3.

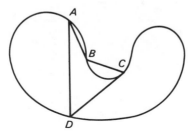

FIGURE 3.26

Two other theorems concern a special property of the length of the boundaries of two convex bodies. The concept of length of a curve is probably a familiar one, but it is restated here.

DEFINITION. The *length* of a simple closed curve is the *least upper bound* of the length of all inscribed polygonal curves. Recall that the least upper bound of a set of real numbers is the smallest real number that is greater than or equal to each number of the set. The length of the boundary of a region is called the *perimeter* of the region.

THEOREM 3.8. If K_1 and K_2 are convex polygonal regions with $K_1 \subseteq K_2$, then the perimeter of K_1 is less than or equal to the perimeter of K_2.

Proof: Let the vertices of K_1 be, in order, $P_0, P_1, \ldots, P_n$, and let l_i, $i = 1, \ldots, n$, be the line through P_i and P_{i-1}. The typical picture is shown in Figure 3.27 to help with this and the following notation. Let S_i be a polygonal region such that $S_i = H_i \cap S_{i-1}$, $i = 1, \ldots, n$, where H_i is the supporting half-plane of K_1 bounded by l_1 and S_0 is the greater region with boundary K_2. In Figure 3.27, $S_1 = H_1 \cap S_0$ is shaded. Since $AB < (AC + CD + DB)$, the perimeter of $S_1 < S_0$. In general,

$$\text{perimeter of } S_1 \leqq \text{perimeter of } S_{i-1}.$$

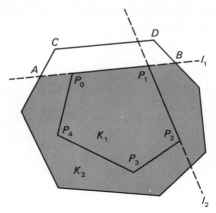

FIGURE 3.27

Since the boundary of S_n is the boundary of K_1, then

$$\text{perimeter of } K_1 \leqq \text{perimeter of } K_2.$$

Theorem 3.8 can be generalized from polygonal regions to convex bodies.

THEOREM 3.9. If K_1 and K_2 are two-dimensional convex bodies with $K_1 \subseteq K_2$, then

$$\text{perimeter of } K_1 \leqq \text{perimeter of } K_2.$$

In Figure 3.28, let A be an interior point of K_1. Each ray with endpoint A intersects the boundaries of K_1 and K_2 in pairs of

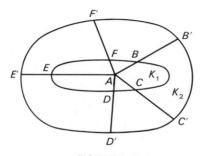

FIGURE 3.28

corresponding points, such as B and B'. There is a one-to-one correspondence between polygons inscribed in K_1 and in K_2. For example, $BCDEF$ is paired with $B'C'D'E'F'$. Since each polygon S_1 of K_1 is contained in the corresponding polygon S_2 of K_2, from Theorem 3.8, the length of S_1 is less than or equal to the length of S_2. The least upper bounds of these sets of lengths are finite, with

$$\text{least upper bound of } S_1 \leq \text{least upper bound of } S_2,$$

so the perimeter of K_1 is less than or equal to the perimeter of K_2.

One important modern application of the theory of convex bodies (convex polygonal regions in this case) is in *linear programming*. The basic idea on which linear programming depends is that the maximum and minimum values of a linear function defined for all points on a convex polygonal region occur at vertices of the polygon. Although a full development of this idea is appropriate to a different mathematics course, a brief explanation of the theory and an example are necessary here in order to make the application meaningful.

To use a very simplified example, suppose that a company plans its production for maximum profit and that they sell just two kinds of items, for different prices. Other limiting restrictions in addition to the selling price must be considered—for example, the number of hours required to manufacture each item and the availability of machines and labor. The problem is to decide how many of each item to manufacture for maximum profit.

From a mathematical point of view, the problem is to find values for x and y that will maximize a linear function of the form $ax + by + c$, where x and y can take on only certain values because of the restrictions. The general theorem on which the application depends is stated without proof:

THEOREM 3.10. If a function $ax + by + c$ is defined for each point of a convex polygonal region, the maximum value occurs for the coordinates of one vertex and the minimum value occurs for the coordinates of another vertex.

EXAMPLE. In order to produce the first kind of item for a profit of \$50, the first machine must be used for one hour and the second machine for two hours. In order to produce the second kind of item for a profit of $40, the first machine must be used for one hour and the second machine for one-half hour. Each machine cannot be operated more than 12 hours a day. How many of each item should be manufactured each day for a maximum profit?

If x represents the number of the first kind of item per day and y represents the number of the second kind of item per day for a maximum profit, then

$$x \geq 0,$$

$$y \geq 0.$$

The inequalities, stating the restrictions on the time for the machines, are

$$x + y \leq 12,$$

$$2x + \tfrac{1}{2}y \leq 12.$$

Each of the four inequalities is represented in Figure 3.29 as a closed half-plane, and the intersection of these four convex sets is the convex polygonal region $ABCD$.

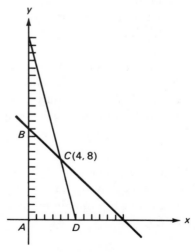

FIGURE 3.29

For the example, the function giving the profit is $f = 50x + 40y$, and the values at the vertices of the polygon are:

$$\text{at } A \ (0,0), \qquad f = 0;$$
$$\text{at } B \ (0,12), \qquad f = 480;$$
$$\text{at } C \ (4,8), \qquad f = 520;$$
$$\text{at } D \ (6,0), \qquad f = 300.$$

For a maximum profit, the company should make four of the first item and eight of the second. The profit will be $520 per day.

EXERCISES   3.3

For Exercises 1–8, indicate which sets are convex bodies. For each set that is not a convex body, explain why not.

1. Triangular region.

2. Line.

3. Angle and its interior.

4. Circle.

5. Half-plane.

6. Nonempty intersection of two convex sets.

7. Ray.

8. Convex polygonal region.

9. Sketch two examples to show that Theorem 3.7 does not necessarily hold if the points P_i are not successive.

10. Complete the proof of Theorem 3.7.

11. Sketch two examples to show that Theorem 3.8 does not necessarily hold if the regions are nonconvex.

12. Describe all possible one-dimensional convex bodies.

13. Show that a plane convex body has a tangent at boundary point A if and only if there is exactly one supporting line for the body at A.

14. Rework the example given for linear programming with the following conditions: for the first item, the first machine must be run two hours and the second machine three hours; the profit on the first item is $90 per item.

15. A rancher has space for 300 cows. He wants no more than 200 of one breed. If he makes a profit of $40 on a Hereford and $50 on a Black Angus, how many of each should he raise for a maximum profit?

3.4   CONVEX BODIES IN THREE-SPACE

Many of the concepts and theorems for convex bodies in three dimensions are the same as or closely analogous to those already discussed for two dimensions in the last section.

DEFINITION. Plane π is called a *supporting plane* of a three-dimensional set S if and only if π contains at least one boundary point of S and S lies entirely in one of the closed half-spaces determined by π. (See Figure 3.30.) The closed half-space determined by π which contains S is called a supporting half-space of S.

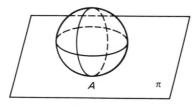

FIGURE 3.30

Things from the physical world that suggest sets and supporting planes are a basketball on a floor and a polar map, which is made by projecting points onto a plane tangent to the globe at the north pole,

(a)   (b)

FIGURE 3.31

as shown in Figure 3.31a. The lines meeting at the pole are lines of longitude and the circles are lines of latitude (Figure 3.31b).

For three-space, a tangent cone is actually the interior of a cone, as in Figure 3.32. There is no direct analogue in three-space to semi-tangents, since the boundary of the tangent is the cone itself. If the boundary cone of the tangent cone for a point A is a plane π, then π is called the *tangent plane* at A.

FIGURE 3.32

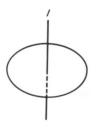

FIGURE 3.33

Supporting lines may also be defined for three-dimensional convex sets, but the definition of supporting lines used for two dimensions will not suffice, since a line does not determine unique half-planes or half-spaces in space. Line l is called a *supporting line* of a three-dimensional convex set K if and only if l contains at least one boundary point but no interior points of K. This definition guarantees that l in Figure 3.33 will not be considered as a supporting line for a two-dimensional set considered in three dimensions, since the definition applies only to three-dimensional sets.

Theorem 3.2 from Section 3.2 corresponds to the following theorem in three-space. The proof is quite similar and is left as an exercise.

THEOREM 3.11. π is a supporting plane of the three-dimensional set K if and only if π contains boundary points but not interior points of K.

The concepts of regular point and corner point can be extended to three-space. A boundary point of a convex set K is regular if and only if K has a tangent plane at the point. The vertices of a convex polyhedral region are examples of corner points, whereas each point on a sphere is a regular point for the spherical region.

The next theorem is the three-space analogue of Theorem 3.3. Before stating the theorem, the concept of surface must be introduced. A *surface* is the graph of a set of equations of the form $x = f(t)$, $y = g(t)$, $z = h(t)$, for f, g, and h continuous functions and the domain of t an interval of real numbers. A *simple closed surface* has a single interior. It partitions every spherical neighborhood into two disjoint sets if the center is a point of the surface.

THEOREM 3.12. A simple closed surface S is the boundary of a convex set K if and only if through each point of S there passes at least one supporting plane of K.

Proof:

a. Proof of the inverse of the only if statement.

Points A and B can be found with A an interior point such that $\overline{AB}$ contains a boundary point C if set K is nonconvex, as in Figure 3.34. No plane through A and B can be a supporting plane,

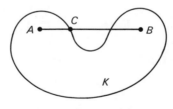

FIGURE 3.34

since A is an interior point. Any other plane through C separates A from B and cannot be a supporting plane. Thus, there is no supporting plane of K through C.

b. The proof of the only if statement depends on the concept of projection, to be explained in Chapter 7. It consists of showing that if the curve is the boundary of a convex set, then there is a supporting plane at each boundary point. This proof is not given.

The next theorem relates two-dimensional and three-dimensional convex bodies.

THEOREM 3.13. Let l be a supporting line of the three-dimensional convex body K at point B, and let π be the plane determined by l and an interior point A of K. Then $\pi \cap K$ is a two-dimensional convex body that has l as a supporting line.

A is also an interior point of $\pi \cap K$, so $\pi \cap K$ is a two-dimensional convex body. But suppose l is not a supporting line of $K \cap \pi$. Since B lies on l and B is a boundary point of $K \cap \pi$, l contains an interior point C of $K \cap \pi$, as in Figure 3.35. There is a point D of $K \cap \pi$ such that C lies between A and D. Since A is an interior point of K and $D \in K$, it follows that C is an interior point of K. This means that l is not a supporting line of K, contrary to the assumption.

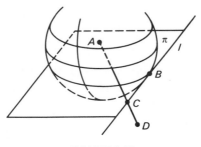

FIGURE 3.35

The last theorem of this section is a three-space counterpart of Theorem 3.4.

THEOREM 3.14. A three-dimensional closed convex set (a proper subset of space) is identical to the intersection of all its supporting half-spaces.

This theorem is proved in a way analogous to the proof of Theorem 3.4, but the proof is left as an exercise.

EXERCISES   3.4

1. Give other examples from the physical world representing supporting planes and three-dimensional sets.

2. Describe the tangent cone for a vertex of a cubical solid.

3. Is every line in a supporting plane for a convex set a supporting line for the set? Why?

Which of the sets of points in Exercises 4–8 are three-dimensional convex bodies?

4. Tetrahedral solid.

5. Sphere.

6. Half-space.

7. Convex polyhedral solid.

8. Cubical solid.

9. Prove Theorem 3.11.

10. Use Theorem 3.14 to give a definition for a tetrahedral solid.

11. Prove Theorem 3.14.

3.5   CONVEX HULLS

The last two sections advanced the theory of convexity by concentrating on properties of convex bodies. In this section, a major new idea is introduced—that of a convex hull.

DEFINITION. The *convex hull* (sometimes called the *convex cover*) of a set of points S is the smallest convex set containing S.

Examples of convex hulls are shown in Figure 3.36. The concepts of convex set and convex hull are further related by the following theorem, whose proof is left as an exercise.

THEOREM 3.15. A set is convex if and only if it is its own convex hull.

In many problems that involve packing objects in convex containers, the convex hull must fit within the container, so the convex

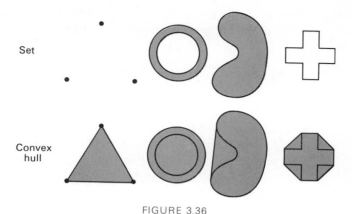

Set

Convex
hull

FIGURE 3.36

hull, rather than the object itself, must be considered. For example, a shoe box is rectangular, rather than the shape of a shoe; a typewriter cover is not usually the shape of a typewriter.

The points on the boundary of a convex set have previously been classified as corner points or regular points. In order to investigate additional relationships between sets and their convex hulls, it is necessary to introduce a second classification based on whether a boundary point is or is not an extreme point.

DEFINITION. A point A of convex set K is called an *extreme point* of K if and only if there are no two points P_1 and P_2 of K such that A is a point of $\overset{\circ}{\overline{P_1 P_2}}\overset{\circ}{}$.

In Figure 3.37, A, B, C, D, E are extreme points, whereas F and G are not. Every point on the boundary of a circular region is an

FIGURE 3.37

extreme point. The connection between extreme points and convex hulls for some sets is made clear in the following two theorems.

THEOREM 3.16. A convex polygonal region S is the convex hull K of its extreme points.

The extreme points are the vertices. Since each pair of extreme points belongs to K, each edge of S joining extreme points belongs to K. Any line through an interior point of S intersects the boundary in two points belonging to K. Then $S \subset K$. Since K is the smallest convex set containing the vertices, $K \subset S$ and therefore $S = K$.

THEOREM 3.17. The convex hull of a finite number of points in a plane is a convex polygonal region. Examples of convex hulls for a finite number of points in a plane are shown in Figure 3.38.

FIGURE 3.38

Let K be the convex hull of the set of points $S = \{P_1, \ldots, P_n\}$ lying in a plane. If K is one dimensional, then K is a line segment that may be considered a special case of a convex polygonal region.

If K is two dimensional, it is bounded and the boundary is a simple closed curve T. Use a typical figure (Figure 3.39) to help with the

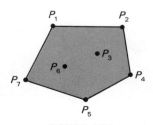

FIGURE 3.39

notation. Let U be the intersection of the finite number of supporting half-planes of S whose boundaries contain at least two points of S. For example, $\overleftrightarrow{P_1 P_2}$ and $\overleftrightarrow{P_2 P_4}$ determine supporting half-planes, whereas $\overleftrightarrow{P_1 P_3}$ and $\overleftrightarrow{P_6 P_7}$ do not.

Since U is convex and $S \subset U$, then $K \subset U$. The rest of the proof involves showing that $U \subset K$.

Suppose, as in Figure 3.40, that $A \notin K$ and that B is the point of K at a minimum distance from A. The line l perpendicular to $\overline{AB}$ is a supporting line of K at B.

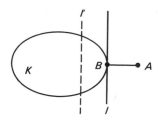

FIGURE 3.40

At least one point, say P_1, of S must lie on l; otherwise l' could be located parallel to l, and there would be a convex proper subset of K on one side of l' that would contain S, contrary to the definition of K as the convex hull.

Now think of line l being rotated about P_1, as in Figure 3.41,

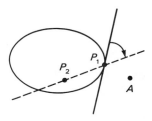

FIGURE 3.41

keeping point A in the same half-plane until a second point P_2 of S is contained in l. The half-plane opposite point A will then be one of the supporting half-planes of U. This means that A, which is not in K, is not in U. The conclusion is that K is a subset of U and, finally, that

$K = U$. The boundary of K consists of the union of line segments, so that K is a convex polygonal region.

The idea of a convex cover can be used in yet another test to determine whether or not a simple closed curve is the boundary of a convex set.

THEOREM 3.18. Let S be an arbitrary finite set of points on a simple closed curve T. T is the boundary of a convex set K if and only if no point of S is an interior point of the convex cover K' of S for all such sets S.

Figure 3.42 helps to make clear the meaning of this theorem. For the convex set in Figure 3.42a, the convex hull of any finite number

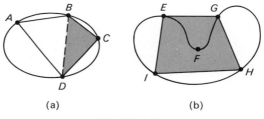

(a)   (b)

FIGURE 3.42

of points, such as B, C, D, does not contain any other points on the boundary. For Figure 3.42b, however, the convex hull for $EGHI$ contains boundary point F.

Proof:

a. If T bounds a convex set K, then, by definition of convex hull, $K' \subseteq K$. Therefore, all interior points of K' are also interior points of K.

b. The proof of the converse consists of showing that if K is not convex, then some member of S is an interior point of the convex hull of S (Exercise 5, Exercise Set 3.5).

The concept of convex hull is easily extended to sets in three dimensions. A convex polyhedral solid is the convex hull of its extreme points, the vertices. Similarly, the convex hull of a finite number of points in space is a convex polyhedral solid.

The final theorem in this section is also a three-space generalization of Theorem 3.18.

THEOREM 3.19. Let S be a finite set of points on a simple closed surface T. T is the boundary of a convex set if and only if no point of S is an interior point of the convex hull U of S for all such sets S.

EXERCISES   3.5

1. Describe or sketch in two dimensions the convex cover for each set listed below:

   a. Circle.
   b. Triangle.
   c. Two intersecting lines.
   d. Two nonintersecting circles.
   e. Parabola.
   f. Hyperbola.
   g. Angle.
   h. Four distinct points, no three of which are collinear.

2. Give examples involving practical applications of packing objects in convex containers.

3. Prove Theorem 3.15.

4. Can an extreme point be:

   a. A regular point?
   b. A corner point?

5. Complete the proof of Theorem 3.18.

6. Give an example of a set of points that is closed, but whose convex hull is not closed.

7. Prove that a convex body is the convex hull of its extreme points.

8. Describe the convex cover of:

   a. A dihedral angle.
   b. Two skew lines.
   c. A sphere.
   d. Four distinct noncoplanar points.

9. Can the convex cover of any two-dimensional set be three dimensional?

10. Prove Theorem 3.19.

11. Prove that every supporting plane of the convex cover of a closed bounded set contains at least one point of the set.

3.6   WIDTH OF A SET

A significant property of sets of points that leads to some modern applications of convexity is the concept of *width* of a set.

DEFINITION. The perpendicular distance along a line *l* between two parallel supporting lines of a two-dimensional set or two parallel supporting planes of a bounded three-dimensional set is the width of the set in the direction indicated by the line *l*.

In Figure 3.43, *w* is the width of the set in the direction of *l*. In general, a set has different widths in different directions. Figure 3.44 shows two examples of width in particular directions for a two-dimensional set. The maximum width of a convex body is called the *diameter*. Note that this agrees with the ordinary concept of diameter of a circular region.

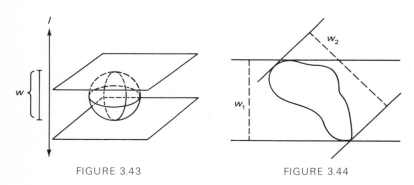

FIGURE 3.43   FIGURE 3.44

For a set that is not a convex body, the concepts of diameter and maximum width are more difficult to relate. The diameter of any set can be defined as the least upper bound of the distances between any two points of the set. Then, for a closed bounded set, it can be proved that the diameter is equal to the maximum width.

Several special properties are associated with the direction of maximum width. Two of these are contained in the following two theorems.

THEOREM 3.20. Let π and π' be parallel supporting planes in a direction of maximum width. If A is any point of $\pi \cap S$, then the line l through A perpendicular to π and π' intersects π' in a point B that is a point of S.

Assume the contrary, as in Figure 3.45. Then there is a point C of $\pi' \cap S$; but $AC > AB$, so that line l is not in a direction of maximum width, contrary to assumption.

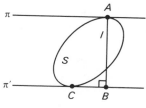

FIGURE 3.45

The proof of the next theorem is left as an exercise (Exercise 11, Exercise Set 3.6).

THEOREM 3.21. Let π and π' be parallel supporting planes for closed set S in a direction of maximum width. Then $\pi \cap S$ and $\pi' \cap S$ each contain exactly one point.

For some special cases, the width of a set is the same for all directions. Sets of this kind are called sets of *constant width*. A set of constant width does not need to be convex, as is shown by the examples in Figure 3.46.

FIGURE 3.46

The most common convex set of constant width is a circular region in two dimensions. Euler, prior to 1800, seems to have been the first to study convex sets of constant width that were not circular regions. It was not until 1953 that Hammer and Sobyzk gave the first satisfactory general account of all such sets.

The simplest example of a convex set of constant width that is not a circular region is shown in Figure 3.47. Let $\triangle ABC$ be an equilateral triangle with the length of each side w. With each vertex as center, construct the shorter arc (of radius w) joining the other two vertices. The union of these three arcs is the boundary of a set of constant width w called a *Reuleaux triangle,* after an early contributor to the study of sets of constant width. Note that a Reuleaux triangle is actually a region, rather than a curve.

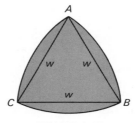

FIGURE 3.47

An interesting difference between the Reuleaux triangle and a circular region is that a Reuleaux triangle has corner points at A, B, and C. In recent years, the mathematical properties of the Reuleaux triangle have led to some extremely important applications. One such application is in the cylinder of a Wankel engine, as shown in Figure 3.48. The Reuleaux triangle inside the double elliptical chamber allows more than one phase of operation of the engine to take place simultaneously. This type engine is being used increasingly in modern automobiles. The same type of efficient engine has been used in snowmobiles. For further information, consult *Scientific American,* February, 1969.

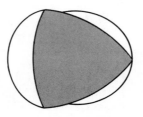

FIGURE 3.48

A second application of a Reuleaux triangle is in the gear for driving a movie film. The action desired is a brief, quick movement, followed by a momentary stoppage. See Figure 3.49. A third application is the construction of a drill that makes a hole with straight sides. This is possible because a Reuleaux triangle can be moved around (but not rotated) inside a square; hence the shape of the drill bit is a Reuleaux triangle, and the bit moves in an eccentric path (Figure 3.50).

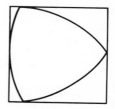

FIGURE 3.49

FIGURE 3.50

Some sets of constant width that are not circular regions also do not have corner points. Let ABC be an equilateral triangle with edge d, and extend each side a distance h past the vertices, as in Figure 3.51. With each vertex as center and with radius h, construct the arcs $A'A''$, $B'B''$, and $C'C''$. Then with each vertex as center and with radius $d + h$, construct the arcs $A''B'$, $B''C'$, and $C''A'$. The six arcs are the boundary of a set of constant width $d + 2h$.

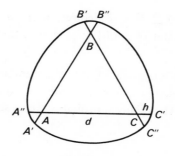

FIGURE 3.51

Sets of constant width of still another type may be constructed in a manner similar to that used for a Reuleaux triangle. Sets of this type use regular polygons with an odd number of sides and are sometimes called *Reuleaux polygons*. A Reuleaux polygon with five sides is shown in Figure 3.52. In general, any set of constant width may be considered as the *orthogonal trajectory* of sets of intersecting lines. The concept of orthogonal trajectory is illustrated in part for a Reuleaux triangle in Figure 3.53. The orthogonal trajectory of a set of lines may be thought of as a curve intersecting each line of the set at right angles. Each ray with endpoint A that intersects $\overset{\frown}{BC}$ does so at right angles. A similar statement may be made for $\overset{\frown}{AC}$ using rays from B and $\overset{\frown}{AB}$ using rays from C.

FIGURE 3.52   FIGURE 3.53

Sets of constant width are still being studied intensively by leading research mathematicians. Several of the simpler properties are given here, but the interested reader may pursue the topic in recent professional journals.

THEOREM 3.22. If K is a convex body of constant width and if $A \notin K$, then the diameter of $K \cup \{A\}$ is greater than the diameter of K.

The intuitive meaning of this theorem is that other outside points cannot be joined to a convex body of constant width without increasing the width. The theorem probably seems self-evident, at least for a circular region.

Proof: In Figure 3.54, suppose K has constant width w and $A \notin K$. Let B be a point of K closest to A and let π be the plane through B perpendicular to $\overleftrightarrow{AB}$. Then π is a supporting plane of K, and $BB' = w$. But then $AB + BB' > BB' = w$. This proof is written for a convex body in three-space, but it can easily be modified for two-space.

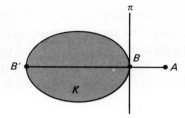

FIGURE 3.54

The converse of this theorem can also be proved, but the proof is not given here. Theorem 3.22 and its converse describe a convex body of constant width as containing all points possible without increasing the diameter.

Consider all convex bodies with the same constant width w. It is easy to find the perimeter (the length of the boundary) of several of these. Thus, for a circular region, the perimeter is πw. Interestingly enough, this is also the perimeter of the Reuleaux triangle with width w. The following theorem, not proved in this text, describes the general situation.

THEOREM 3.23. *Barbier's theorem.* The perimeter of a plane convex body of constant width w is πw.

Certainly, all convex sets with the same constant width do not have the same measure of area. The set with maximum area is the circular region. The set with minimum area is named in the following theorem.

THEOREM 3.24. *Blaschke-Lebesque theorem.* The set of constant width with the least area is the Reuleaux triangle.

The Reuleaux triangle has some properties of a circular region, but it does not have a center. In fact, the only convex bodies of constant width having a center are circular regions in the plane and spherical regions in three dimensions.

One of the current areas of investigation in convexity is the concept of universal cover. A *universal cover* is defined as a plane region that can be used to cover every set whose diameter is one. In other words, any set with a diameter of one can be located in the plane in such a way that it is a subset of a universal cover.

A set that will cover every convex set of diameter one will cover every set of diameter one. The smallest square universal cover is a unit square. The smallest equilateral triangle that is a universal cover has an incircle of diameter one. The general problem of the smallest universal cover of any given shape has not been completely solved.

EXERCISES   3.6

Find the minimum width and the diameter of each set in Exercises 1–4.

1. Square region with edge 1 inch.

2. Rectangular region 2 inches by 3 inches.

3. Reuleaux triangle constructed on equilateral triangle with edge 1 inch.

4. Isosceles triangle with edges 7 inches and noncongruent edge 3 inches.

5. Picture other nonconvex sets of constant width.

6. Draw a Reuleaux polygon of seven sides.

7. Construct a set of constant width as in Figure 3.51, but begin with a pentagon instead of a triangle.

8. Show that the perimeter of a Reuleaux triangle is πw.

9. Find the maximum and minimum areas of plane sets of constant width four.

10. Find the length of the edge of the smallest equilateral triangle that is a universal cover.

11. Prove Theorem 3.21.

12. Give an example to show that Theorem 3.22 is not necessarily true for a nonconvex body of constant width.

13. Give an example to show that Theorem 3.22 is not necessarily true for a convex body *not* of constant width.

3.7   HELLY'S THEOREM AND
APPLICATIONS

The central theorem of this section is named for the Austrian mathematician Eduard Helly (1884–1943). Helly studied at the University of Vienna and at Gottingen. His theorem was discovered in 1913 and published in 1923. Interestingly enough, Helly, like Poncelet, who discovered projective geometry, spent several years as a prisoner of the Russians. In 1938, he and his family moved to the United States. The theorem and the wealth of related material constitute a substantial portion of the recent discoveries in the geometry of convexity.

THEOREM 3.25. *Helly's theorem.* Let $K = \{K_1, K_2, \ldots, K_N\}$ be N convex sets of points, $N \geqq n + 1$, lying in n-space, $n = 1, 2,$ or 3, so that every $n + 1$ sets have a nonempty intersection. Then the intersection of all the sets is not empty.

Figure 3.55 is an illustration of Helly's theorem for four convex sets in a plane.

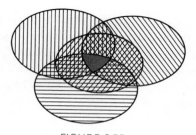

FIGURE 3.55

The proof is given for three dimensions. For $N = 4$, the theorem is trivially true. The proof can be completed by induction on N. If necessary, the reader should review the idea of mathematical induction before continuing the proof. Assume the theorem true for any $4, \ldots, m$ convex sets of points and let $K_1, \ldots, K_m, K_{m+1}$ be convex sets, any four of which have a common point. Let $K = K_m \cap K_{m+1}$. Since the theorem is assumed to be true for $N = 5$, K_m, K_{m+1}, and any other three sets K_i, K_j, K_k have a point in common. Therefore, $\{K_1, K_2, \ldots, K_{m-1}, K\}$ is a set of m convex sets, each four of which have a point in common. By the induction assumption, the intersection of all these sets is not empty. This common point is also common to $K_1, \ldots, K_m, K_{m+1}$, since each point of K is a point of both K_m and K_{m+1}. The theorem holds for $m + 1$ sets when it holds for up through m sets, and the theorem is proved by induction.

Many applications of Helly's theorem depend on a more general form for a nondenumerable infinity of sets, stated here without proof.

THEOREM 3.26. For any collection of convex bodies in n-space, if every $n + 1$ of them have a common point, then all the convex bodies have a common point.

One of the interesting applications of Helly's theorem shows the existence of points that behave somewhat like the center of symmetry of a set of points even when the set of points is not symmetric.

THEOREM 3.27. Let $S = \{P_1, \ldots, P_n\}$ be any finite set of points in space. Then there is a point A such that every closed half-space formed by a plane through A contains at least $n/4$ points of S.

As an illustration, suppose S consists of six points in space. The theorem says that there is a point A such that every closed half-space formed by a plane through A contains at least $6/4$ points of S. This means each half-plane must actually contain at least two points of S, since it contains more than one. It is also important to point out that A does not have to be a member of S.

Proof: Since S is a finite set of points, there exists a spherical region B containing all the points P_i. Consider the set of all closed half-spaces containing more than $3n/4$ points of S. Let H_1, H_2, H_3, H_4 be any four of these closed half-spaces. In the study of set theory, it is proved that

$$(H_1 \cap H_2 \cap H_3 \cap H_4)' = H_1' \cup H_2' \cup H_3' \cup H_4',$$

where H_i' is the complement of H_i. Since each H_i' contains less than $n/4$ points of S, $H_1' \cup H_2' \cup H_3' \cup H_4'$ does not contain all n of the points of S. Then $H_1 \cap H_2 \cap H_3 \cap H_4$ must contain at least one point of S.

Since every four of the closed half-spaces have a common point of S, Helly's theorem (in the form of Theorem 3.26) can be applied to the intersections of these half-spaces with B to conclude there is a point A common to all such half-spaces.

This point A is the desired point in the theorem. If it were not, then there would be a plane π through A that is the boundary of a closed half-space containing less than $n/4$ points of S. The opposite open half-space H will contain more than $3n/4$ points of S. Let π' be the plane parallel to π passing through the points of $S \cap H$ closest to π. The closed half-space H' with boundary π' and lying in H contains more than $3n/4$ points of S, but H' does not contain A. This is a contradiction, since A was assumed to be in the intersection of all half-spaces containing more than $3n/4$ of the points. It should be obvious that the theorem does not explain how the point A is actually found.

The applications of Helly's theorem are surprisingly numerous; only a few are listed here. A theorem similar to Theorem 3.27, but involving volume, is the following, stated without proof.

THEOREM 3.28. Let S be a bounded set of points in space having volume V. Then there is a point A such that every closed half-space formed by a plane through A intersects S in a set with volume at least $V/4$.

Helly's theorem is rather closely related to other geometric concepts, including the ideas of universal cover and width of a set, as is evident by the final four applications. Attempting to draw a figure to illustrate the last three will lead to additional understanding.

THEOREM 3.29. If each three of n points in a plane can be enclosed in a circle of radius one, then all n points can be enclosed in such a circle.

The proof consists of showing that a point (the center of the desired circle) exists that is not more than one unit from each of the points. A typical figure is shown in Figure 3.56.

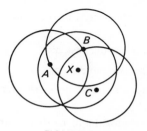

FIGURE 3.56

Let a circle with center X be a unit circle enclosing three of the n points, A, B, C. The three circles with centers A, B, C all contain X.

By Helly's theorem, since any three of the unit circles with the given points as center have a nonempty intersection, all of the circles have a nonempty intersection. A point in this common intersection will serve as the center of a unit circle enclosing all the points.

THEOREM 3.30. *Jung's theorem.* For n points of the plane, such that each two are less than one unit apart, all can be enclosed in a circle of radius $\sqrt{3}/3$.

THEOREM 3.31. *Blaschke's theorem.* Every bounded convex figure of constant width one contains a circle of radius $1/3$.

THEOREM 3.32. Given n parallel line segments in the same plane, if there exists a line that intersects each three of them, then there is a line intersecting all the line segments.

The geometry of convexity is a rapidly expanding area of modern geometry. Many of the familiar figures of Euclidean geometry have been shown to possess new properties when they are considered as convex sets. The chapter is deliberately open-ended in the hope that the reader will want to seek recent research articles on convexity.

EXERCISES   3.7

1. Draw a picture with six convex sets in a plane to illustrate Helly's theorem.
2. Give an example in one dimension to show that Helly's theorem does not hold for nonconvex sets.
3. Give an example in two dimensions to show that Helly's theorem does not hold for nonconvex sets.
4. Prove Helly's theorem for one dimension.
5. Prove Helly's theorem for two dimensions.
6. For the eight vertices of a cube, does Theorem 3.27 apply for A the center of the cube?
7. State and prove a theorem for the plane that is analogous to Theorem 3.27.
8. Verify Theorem 3.31 for a Reuleaux triangle of constant width one.
9. Draw a picture to verify Theorem 3.32.
10. Prove Theorem 3.28.

CHAPTER 4

EUCLIDEAN GEOMETRY OF THE POLYGON AND CIRCLE

4.1   FUNDAMENTAL CONCEPTS AND THEOREMS

In Chapters 2 and 3, much of the material of Euclidean geometry was studied from two different points of view. Chapter 2 emphasized the idea of Euclidean geometry as a study of the invariant properties of sets of points under Euclidean or similarity transformations. Chapter 3 concentrated on the concept of convexity, one of the properties preserved by similarity transformations (as well as by somewhat more general transformations). The present chapter begins with a survey of some fundamental concepts and theorems concerning polygons (particularly triangles) and circles—concepts and theorems that have long been a part of traditional Euclidean geometry. Then it progresses rapidly to the wealth of modern material that has been discovered since 1800 about these basic figures of synthetic Euclidean geometry (geometry that does not use coordinates). The last section includes three significant applications of the Euclidean geometry of the polygon and circle—the golden ratio, tessellations, and caroms.

It is essential that the student of modern college geometry understand several useful key concepts and theorems from elementary geometry. Among these are the concurrence theorems, sometimes found in high school geometry texts, that identify four significant points connected with a triangle.

THEOREM 4.1. The perpendicular bisectors of the sides of a triangle are concurrent at a point called the *circumcenter*.

Theorem 4.1 is illustrated in Figure 4.1. It is significant when any *three* lines have a common point in a geometry, because any three lines generally intersect by pairs to determine three distinct points,

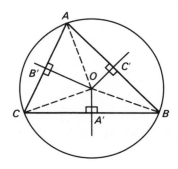

FIGURE 4.1

rather than a single one. The proof of Theorem 4.1 depends on showing that the point of intersection of two of the perpendicular bisectors, say $\overleftrightarrow{B'O}$ and $\overleftrightarrow{C'O}$ in Figure 4.1, also lies on the perpendicular bisector of the third side. Point O is equidistant from A and C because it is on $\overleftrightarrow{B'O}$; it is also equidistant from A and B because it is on $\overleftrightarrow{C'O}$. Therefore, O is equidistant from B and C on the perpendicular bisector of $\overline{BC}$. Since O is the point in a triangle equidistant from the three vertices, it is called the circumcenter to show that it is the center of the *circumcircle,* the unique circle containing all three vertices of a triangle.

In Figure 4.2, triangle $A'B'C'$ is formed by joining the midpoints of the sides of the original triangle. Since the segments joining the midpoints of two sides of a triangle are parallel to the third side, the perpendicular bisectors of the sides of triangle ABC are also perpendicular to the sides of triangle $A'B'C'$. This means that $\overleftrightarrow{A'O}$ is perpendicular to

FIGURE 4.2

$\overleftrightarrow{B'C'}$ and so on; thus, the perpendicular bisectors of the sides of triangle ABC are the altitudes of triangle $A'B'C'$. Reversing this statement leads to the claim that the altitudes of a triangle are the perpendicular bisectors of the sides of another triangle and hence are concurrent (by Theorem 4.1).

THEOREM 4.2. The altitudes of a triangle are concurrent at a point called the *orthocenter.*

Figure 4.3 shows triangle ABC and its orthocenter H. The four

FIGURE 4.3

points A, B, C, and H constitute an *orthocentric set* of four points, so named because each of the four points is the orthocenter of the triangle formed by the other three.

Two additional theorems about concurrency involve the internal angle bisectors and the medians of a triangle.

THEOREM 4.3. The internal bisectors of the angles of a triangle meet at a point called the *incenter*.

The proof of Theorem 4.3 (the details are left as Exercise 7, Exercise Set 4.1) depends on the fact that every point on the internal bisector of an angle is equidistant from the adjacent sides of the angle. For example, in Figure 4.4a, if I is on the angle bisector of angle B, then $\overline{IY} \cong \overline{IX}$. Since I is equidistant from all three sides of the triangle, it is the center of the *incircle*, a circle inscribed in the triangle. This means that the three sides of the triangle are tangent to the incircle.

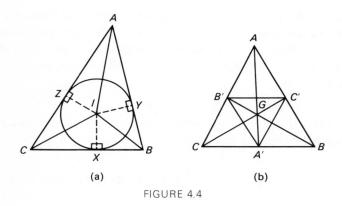

(a)   (b)

FIGURE 4.4

THEOREM 4.4. The medians of a triangle meet at a point called the *centroid*.

The centroid is the *center of gravity* for a triangle. Recall that medians join the vertex and the midpoint of the opposite side of a triangle. In Figure 4.4b, triangles CBG and $GC'B'$ are similar with a ratio of similarity of two to one. The completion of the details of the proof is left as an exercise.

So far, four points of concurrency have been introduced. It is natural to wonder whether the four new points constitute a significant set of points in their own right. The matter will be discussed in Section 4.3, but you may profit from conjecturing about the location of these points now.

Much of the study of the triangle in Euclidean geometry involves work with proportions, and ordinarily this concept is related to similar triangles. A basic proportion used as a tool in college geometry is the property that is connected with internal angle bisectors of the angles of a triangle, stated as Theorem 4.5.

THEOREM 4.5. The internal bisectors of an angle of a triangle divide the opposite side into two segments proportional to the adjacent sides of the triangle.

In Figure 4.5, assume that $\overline{AD}$ is the internal angle bisector of angle A of triangle ABC and that $\overline{CE}$ is parallel to $\overline{AD}$. Because of the parallelism, $\angle ECA \cong \angle CAD \cong \angle CEA$. This means that triangle ECA is isosceles and that $\overline{EA} \cong \overline{AC}$. Now think of the figure as formed by two transversals from B intersecting a pair of parallel lines, $\overleftrightarrow{EC}$ and $\overleftrightarrow{AD}$, so that the following is a true proportion:

$$\frac{CD}{EA} = \frac{DB}{AB} \quad \text{or} \quad \frac{CD}{CA} = \frac{DB}{AB},$$

so that

$$\frac{CD}{DB} = \frac{CA}{AB},$$

as was to be established. Note that directed segments are not employed in this development.

FIGURE 4.5

The proof of the analogous theorem for *external bisectors* is left as Exercise 10, Exercise Set 4.1. It sometimes comes as a surprise to students of geometry that the external bisectors of angles of a triangle have properties corresponding to those of the internal bisectors. Corresponding to the concept of incenter, for example, is the concept of excenter given in the statement of Theorem 4.6.

THEOREM 4.6. The external bisectors of two angles of a triangle meet the internal bisector of the third angle at a point called the *excenter*.

Of course, when talking about external bisectors of a triangle, one must think of the sides as "extended" or consider the sides as lines rather than segments. In Figure 4.6, assume that E is the excenter. It becomes apparent in the proof of Theorem 4.6, left as an exercise, that E is equidistant from all three sides of the triangle; hence E is the center of a circle externally tangent to all three sides of the triangle. This circle is one of the *excircles* of the given triangle.

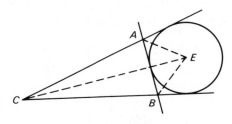

FIGURE 4.6

One other property of internal and external angle bisectors—the fact that they are perpendicular (Exercise 12)—is needed to help prove the following theorem.

THEOREM 4.7. If the distances from a point P to two fixed points have a given ratio, then the set of all locations for the point P is a circle, called the *circle of Apollonius*.

The circle of Apollonius is named after the Greek mathematician Apollonius, who wrote a comprehensive treatment on conic sections prior to 200 B.C. Assume, using the notation of Figure 4.7a, that $PA/PB = c$, a given constant. It must be shown that the set of all locations for point P is a circle. There are two points on $\overleftrightarrow{AB}$ whose distances from A and B are the correct ratio without regard to directed distances. These are indicated by points C and D in the figure. Then, in triangle APB, $\overline{PC}$ and $\overline{PD}$ are internal and external angle bisectors of the angle at P. This can be shown for point C, for example, since

$$\frac{AC}{CB} = \frac{PA}{PB}.$$

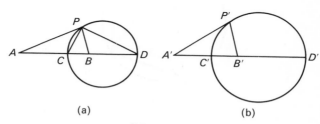

(a)   (b)

FIGURE 4.7

But the internal and external angle bisectors at a vertex of a triangle are perpendicular, so $\angle CPD$ is a right angle. Triangle CPD is inscribed in a semicircle, so P lies on a circle with $\overline{CD}$ as diameter.

Figure 4.7b shows a specific example of the circle of Apollonius, for which $P'A'/P'B' = 5/3$. Each point on the circle is 5/3 as far from A' as it is from B', and the points on the circle are the only points in the plane with this property.

In the proof of Theorem 4.7 and elsewhere, note that even though parallelism (perpendicularity) is a relationship usually defined for lines, use of the concept is extended to segments and rays without difficulty. For example, although it is technically imprecise to say that $\overline{AB}$ is parallel (perpendicular) to $\overline{CD}$, that expression is understood to mean that $\overleftrightarrow{AB}$, of which $\overline{AB}$ is a subset, is parallel (perpendicular) to $\overleftrightarrow{CD}$, of which $\overline{CD}$ is a subset.

Two final theorems about the segments related to a circle are somewhat connected to the previous theorem and also are useful in proving more advanced theorems.

THEOREM 4.8. If a quadrilateral is inscribed in a circle, then the opposite angles are supplementary.

The term *cyclic quadrilateral* is also used for a quadrilateral inscribed in a circle. Study Figure 4.8. You should recall that the measure of an inscribed angle in a circle is half that of its intercepted arc. For example, the measure of angle B is half that of arc ADC. Since angles B and D together intercept arcs with a total measurement of 2π degrees, the angles are supplementary since the sum of their measures is π.

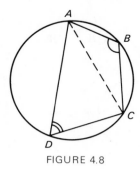

FIGURE 4.8

Opposite sides of an inscribed quadrilateral are sometimes called *antiparallel* with respect to the remaining pair of sides. The prefix "anti" suggests across from or opposite. In Figure 4.8, if angles D and A were supplementary, then $\overline{DC}$ and $\overline{AB}$ would be parallel. But instead, it is the angle opposite or across from A that is supplementary to D, hence the segments are antiparallel instead of parallel.

THEOREM 4.9. The product of the lengths of the segments from an exterior point to the points of intersection of a secant with a circle is equal to the square of the length of the tangent from the point to the circle.

Using the notation of Figure 4.9, the theorem says that

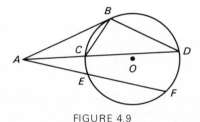

FIGURE 4.9

$AC \cdot AD = (AB)^2$. Recall again that the notation AC indicates a number, the measure of $\overline{AC}$. Triangles ABC and ADB are similar because corresponding angles are congruent; therefore, $AC/AB = AB/AD$, from which the theorem follows.

EXERCISES 4.1

1. Use the notation of Figure 4.3 to name all of the triangles whose vertices are three of the given points and whose orthocenter is the fourth point.

2. Where is the orthocenter of a right triangle?

3. Under what conditions would the orthocenter of a triangle lie outside the triangular region?

4. How many excenters does a triangle have?

5. Could antiparallel segments also be parallel?

6. Prove that the segment joining the midpoints of two sides of a triangle is parallel to the third side.

7. Complete the proof of Theorem 4.3.

8. Prove that the angle between the segments from the incenter to two vertices of a triangle has a measure equal to 90 plus one-half the measure of the angle of the triangle at the third vertex.

9. Complete the proof of Theorem 4.4.

10. Prove that the external bisector of an angle of a triangle divides the opposite side (externally) into two segments proportional to the adjacent sides of the triangle.

11. Prove Theorem 4.6.

12. Prove that the internal and external angle bisectors at a vertex of a triangle are perpendicular.

13. Prove the converse of Theorem 4.8.

14. Prove that two vertices of a triangle and the feet of the altitudes to the sides adjacent to the third vertex can be inscribed in a circle. (The feet are the points of intersection of the altitude with the opposite side of the triangle.)

4.2 SOME THEOREMS LEADING TO MODERN SYNTHETIC GEOMETRY

In surveying the large body of geometric material developed prior to the modern era, it becomes obvious that some of the earlier

theorems are more important than others in leading to later developments. Several of these still-significant ideas are discussed in this section.

Two theorems usually studied together, though proved in different
ages, have great significance because they concern only collinearity and
concurrence and hence are of value in the study of projective geometry
when distance is no longer an invariant. The first of these, Menelaus'
theorem, is credited to the Greek mathematician Menelaus, who lived in
Alexandria about 100 B.C.

THEOREM 4.10. *Menelaus' theorem.* If three points, one on each
side of a triangle (extended if necessary) are collinear, then the product
of the ratios of division of the sides by the points is negative one
(-1), if internal ratios are considered positive and external ratios are
considered negative.

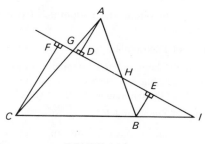

FIGURE 4.10

Suppose, using the notation of Figure 4.10, that points G, H, I
are the three given collinear points, one on each side of the triangle
ABC. The following pairs of similar triangles can be observed:

$$\triangle ADG \sim \triangle CFG,$$

$$\triangle BEI \sim \triangle CFI,$$

$$\triangle BEH \sim \triangle ADH.$$

From the three pairs of similar triangles are derived the three proportions, since pairs of corresponding sides of similar triangles have the
same ratio. Also recall that the notation AG indicates the measure of $\overline{AG}$,
which is a number.

$$\frac{AG}{GC} = \frac{AD}{CF} \qquad \frac{CI}{IB} = -\frac{CF}{BE} \quad \begin{array}{l} \text{(considered negative} \\ \text{because } I \text{ divides} \\ CB \text{ externally)} \end{array} \qquad \frac{BH}{HA} = \frac{BE}{AD}$$

Then

$$\frac{AG}{GC} \cdot \frac{CI}{IB} \cdot \frac{BH}{HA} = -\frac{AD}{CF} \cdot \frac{CF}{BE} \cdot \frac{BE}{AD} = -1.$$

Note that the three ratios may be named by starting at a vertex (say A) and proceeding around the triangle (say in a counterclockwise direction), going from a vertex to a point of division, then to the next vertex, then to the next point of division, and so on. This same pattern may be used when all three points of division are external, as in Figure 4.11:

$$\frac{AE}{AC} \cdot \frac{CF}{FB} \cdot \frac{BD}{DA} = -1.$$

FIGURE 4.11

In case the line containing the three points goes through a vertex, the statement of the theorem can be modified to avoid a zero in the denominator (see Exercise 3, Exercise Set 4.2).

The converse of Menelaus' theorem provides an important tool for proving that three points are collinear.

THEOREM 4.11. *Converse of Menelaus' theorem.* If the product of the ratios of division of the sides by three points, one on each side of a triangle, extended if necessary, is negative one, then the three points are collinear.

Suppose, using the notation of Figure 4.11, that D, E, F are three points on each side of the triangle and that the product of the ratios of division is negative one. Suppose, furthermore, that $\overleftrightarrow{EF}$ meets $\overleftrightarrow{AB}$ at a distinct point D'. It must be shown that $D' = D$. Using Menelaus' theorem for E, F, and D',

$$\frac{AE}{AC} \cdot \frac{CF}{FB} \cdot \frac{BD'}{D'A} = -1.$$

But this means

$$\frac{AE}{AC} \cdot \frac{CF}{FB} \cdot \frac{BD'}{D'A} = \frac{AE}{AC} \cdot \frac{CF}{FB} \cdot \frac{BD}{DA},$$

or

$$\frac{BD'}{D'A} = \frac{BD}{DA}.$$

Since $BA = BD' + D'A$ or $BD + DA$,

$$\frac{BD' + D'A}{D'A} = \frac{BD + DA}{DA}$$

and

$$D'A = DA;$$

therefore, D and D' are the same point, which means that D, E, F are collinear.

Theorem 4.11 can be used to prove several other theorems, one of which is stated below but proved as an exercise.

THEOREM 4.12. The internal bisectors of two angles of a triangle and the external bisector of the third angle intersect the opposite sides of the triangle in three collinear points.

The theorem often paired with the theorem of Menelaus, Ceva's theorem, was discovered about 1678 by the Italian mathematician Ceva. Menelaus' theorem deals with three collinear points, whereas the theorem of Ceva concerns three concurrent lines.

THEOREM 4.13. *Ceva's theorem and its converse.* Three lines that join three points, one on each side of a triangle, to the opposite vertices are concurrent if and only if the product of the ratios of division of the sides is one.

The proof of the direct theorem is given here. The form of the theorem can be written with equal products rather than ratios to avoid zero in the denominator if one of the three points coincides with a vertex. Using the notation of Figure 4.12, Menelaus' theorem applied to triangle ABD and $\overleftrightarrow{CF}$ shows that

$$\frac{AG}{GD} \cdot \frac{DC}{CB} \cdot \frac{BF}{FA} = -1.$$

FIGURE 4.12

For triangle ACD and $\overleftrightarrow{BE}$,

$$\frac{AE}{EC} \cdot \frac{CB}{BD} \cdot \frac{DG}{GA} = -1.$$

Multiplying left and right members of these equations results in

$$\frac{AG}{GD} \cdot \frac{DC}{CB} \cdot \frac{BF}{FA} \cdot \frac{AE}{EC} \cdot \frac{CB}{BD} \cdot \frac{DG}{GA} = 1$$

or

$$\frac{AE}{EC} \cdot \frac{CD}{DB} \cdot \frac{BF}{FA} = 1.$$

Note that the three ratios can be written quickly by following the same pattern described for the theorem of Menelaus: start at a vertex, go to the point of division, then the next vertex, and so on. It should also be noted that similar triangles may be used to prove Ceva's theorem directly, without using the theorem of Menelaus (Exercise 8, Exercise Set 4.2).

The converse of Ceva's theorem, proved indirectly in a way similar to Theorem 4.11, is a useful tool for proving three lines concurrent. For example, it can now be used to give very simple proofs that the medians and the internal bisectors of the angles of a triangle are concurrent (see Exercises 9–11, Exercise Set 4.2).

A more modern application of the converse of Ceva's theorem in ordinary Euclidean geometry is in the proof of the following theorem:

THEOREM 4.14. The segments from the vertices of a triangle to the points of tangency of the incircle are concurrent.

This theorem was proved in the early nineteenth century by J. D. Gergonne, a French mathematician, and the point of concurrency is called the *Gergonne point*.

Using the notation of Figure 4.13, since the two tangents from a point to a circle are congruent, $AE = AF$, $CE = CD$, $BD = BF$, and

$$\frac{AE}{EC} \cdot \frac{CD}{DB} \cdot \frac{BF}{FA} = 1,$$

so the three segments are concurrent by the converse of Ceva's theorem.

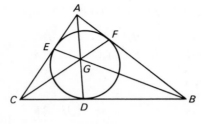

FIGURE 4.13

Many other theorems about concurrency and collinearity have interested geometers throughout the history of geometry. One additional example, credited to the English mathematician Robert Simson (1687–1768), refers to a triangle and a point on its circumcircle.

THEOREM 4.15. The three perpendiculars from a point on the circumcircle to the sides of a given triangle intersect the sides in three collinear points. The line on which the three points lie is called the *Simson line.*

Use the notation of Figure 4.14, with P the point on the circumcircle and D, E, F the feet of the perpendiculars. By the converse of Theorem 4.8, points P, D, A, E are cyclic (lie on a circle). Similarly, P, A, C, B are on another circle, and P, D, B, F are on yet a third circle.

In circle P, D, A, E $\qquad \angle PDE \cong \angle PAE$

$\qquad\qquad\qquad\qquad\quad \angle PAE \cong \angle PAC$

In circle P, A, C, B $\qquad \angle PAC \cong \angle PBC$

$\qquad\qquad\qquad\qquad\quad \angle PBC \cong \angle PBF$

In circle P, D, B, F $\qquad \angle PBF \cong \angle PDF$

Since this establishes the fact that $\angle PDE \cong \angle PDF$, points D, E, F are collinear.

FIGURE 4.14

For a given point P on the circumcircle, one Simson line is determined. The special case when P coincides with a vertex is considered in the exercises.

EXERCISES   4.2

For Exercises 1 and 2, use Figure 4.15. For the triangle and the transversal named in the exercise, write the product of the three ratios equal to negative one by the theorem of Menelaus.

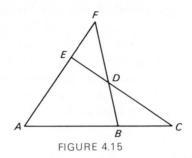

FIGURE 4.15

1.   Triangle ABF and transversal E, D, C.

2.   Triangle BCD and transversal A, E, F.

3.   Write an alternative form of the theorem of Menelaus showing the product of three segments equal to the negative of the product of three other segments. What is the value of the product if one of the points of intersection coincides with a vertex of the triangle?

4.   Prove Theorem 4.12.

5.   Use the converse of Menelaus' theorem to prove that the external bisectors of the angles of a triangle meet the opposite sides of the triangle in three collinear points.

6.   Altogether there are six bisectors of the angles of a triangle, three internal and three external. Prove that these six bisectors meet the opposite sides in six points that are on four lines, three on each line. (Compare this with a finite geometry studied in Chapter 1.)

7.   In Figure 4.16, let A, B, C be the vertices of the triangle, with P the

FIGURE 4.16

point of concurrency of three segments from the vertices. Use Ceva's theorem to write three ratios whose product is equal to one.

8. Prove Ceva's theorem directly, without using the theorem of Menelaus.

9. Use Theorem 4.13 to prove that the medians of a triangle are concurrent.

10. Use Theorem 4.13 to prove that the internal bisectors of the angles of a triangle are concurrent.

11. Use Theorem 4.13 to prove that the external bisectors of two angles of a triangle and the internal bisectors of the third angle are concurrent.

12. Describe the location of the Simson line if the point on the circumcircle is a vertex of the triangle.

4.3 THE NINE-POINT CIRCLE AND EARLY NINETEENTH CENTURY SYNTHETIC GEOMETRY

Renewed interest in dramatic extensions of the classical geometry of the triangle and the circle began in the early nineteenth century. Probably the most significant of the advances at this time was the concept of the nine-point circle, sometimes credited to the German mathematician Feuerbach in 1822.

THEOREM 4.16. The midpoints of the sides of a triangle, the points of intersection of the altitudes and the sides, and the midpoints of the segments joining the orthocenter and the vertices all lie on a circle called the *nine-point circle*.

A nine-point circle is shown in Figure 4.17. In general, the circle intersects each side of the triangle in two distinct points.

FIGURE 4.17

The proof of the theorem follows this pattern:

    a. There is a circle through A', B', C', the midpoints of the sides.

    b. Show that points D, E, F are on this same circle.

    c. Show that points G, H, I are on this same circle.

To prove that D is on the same circle as A', B', C', it is possible to show that $DB'C'A'$ is an isosceles trapezoid and hence, inscribed in the circle. $\overline{A'C'} \cong \overline{DB'}$ because $\overline{A'C'}$ connects the midpoints of the sides of triangle ABC and therefore has a measure equal to half the measure of the base and $\overline{DB'}$ connects the vertex and the midpoint of the hypotenuse of right triangle ADC, which implies that DB is half the hypotenuse.

Now consider the circle with $\overline{IA'}$ as diameter. Point B' is on this circle, since angle $IB'A'$ is a right angle; point C' is also on the circle. Therefore I lies on the same circle as $A'B'C'$.

There is a wealth of additional information about the nine-point circle and its relationships with other sets of points in Euclidean geometry. One of the important ideas is the location of the center of the nine-point circle in reference to several other points previously mentioned. This information depends on a theorem established by the German mathematician Euler in 1765.

THEOREM 4.17. The centroid of a triangle trisects the segment joining the circumcenter and the orthocenter.

The line containing the three points is called the *Euler line*. "Trisect" as used here means that the distance along the line from the circumcenter to the centroid is one-third of the distance along the same line from the circumcenter to the orthocenter.

In Figure 4.18, let O be the circumcenter, G the centroid, and H the orthocenter. The measure of $\overline{AH}$ is twice that of $\overline{OA'}$. This is true because triangles CBI and COA' are similar with a ratio of similarity of 2 to 1, and $IB = AH$ because $AHBI$ is a parallelogram. Now triangles GOA' and GHA are similar with a ratio of 1 to 2; therefore $OG = \frac{1}{2}GH$, and G trisects $\overline{OH}$.

It may be surprising to find that the Euler line also contains the center of the nine-point circle, as indicated in the following theorem.

FIGURE 4.18

THEOREM 4.18. The center of the nine-point circle is the mid-point of the segment whose endpoints are the orthocenter and the circumcenter of a triangle. (See Figure 4.19.)

From Theorem 4.17, $\overline{EH} \cong \overline{OA'}$, and the two segments are also parallel. If O' is the center of the nine-point circle, it is the midpoint of the diameter $\overline{A'E}$. But $OA'HE$ is a parallelogram, and the diagonals of a parallelogram bisect each other. This means that O' is the midpoint of $\overline{OH}$.

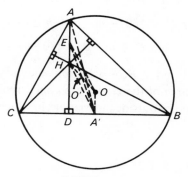

FIGURE 4.19

All the theorems about concurrence considered so far relate to the concurrence of three lines. *Miquel's theorem*, proved in 1838, is significant in part because it considers the concurrence of sets of three circles associated with any triangle.

THEOREM 4.19. If three points are chosen, one on each side of a triangle, then the three circles determined by a vertex and the two points on the adjacent sides meet at a point called the *Miquel point*.

Using the notation of Figure 4.20, let *D, E, F* be the arbitrary points on the sides of triangle *ABC*. Suppose that circles with centers

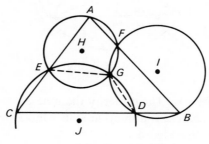

FIGURE 4.20

I and *J* intersect at point *G*. In circle *J*, $\angle EGD$ and $\angle ECD$ are supplementary, and in circle *I*, $\angle FGD$ and $\angle DBA$ are supplementary. In the rest of the proof, the notation *m* means the measure of the angle. Since

$$m\angle EGD + m\angle DGF + m\angle EGF = 360,$$

$$180 - m\angle C + 180 - m\angle B + m\angle EGF = 360,$$

$$m\angle EGF = m\angle C + m\angle B = 180 - m\angle A.$$

This means that $\angle A$ and $\angle EGF$ are supplementary, so *A, F, G, E* are on a circle and all three of the circles are concurrent. It is possible that the Miquel point could be outside the triangle, in which case the proof must be modified slightly.

EXERCISES   4.3

1. A triangle has how many:
    a. Nine-point circles?   b. Euler lines?   c. Miquel points?

In the proof of Theorem 4.16 (for Exercises 2–4),

2. Prove that $DB'C'A'$ is an isosceles trapezoid.

3. Prove that the segment connecting the vertex of the right angle of a right triangle with the midpoint of the hypotenuse has a measure half that of the hypotenuse.

4. Prove that $IB'A'$ is a right angle.

5. Complete the details of the proof of Theorem 4.17 not included in the text.

6. Prove that the radius of the nine-point circle is half that of the circumcircle.

7. Prove that the four triangles formed by the points of an orthocentric group of points have the same nine-point circle.

8. a. Draw a figure showing an example in which the Miquel point is outside the triangle.
    b. Modify the proof as necessary so that the three circles are concurrent.

9. Show that, on the Euler line, the centroid and the orthocenter divide internally and externally in the same ratio the segment whose endpoints are the circumcenter and the center of the nine-point circle.

10. Prove that the Miquel point is a point on the circumcircle if the three points on the sides of the triangle are collinear.

11. Prove that if the Miquel point is on the circumcircle, then the three points on the sides of the triangle are collinear.

4.4 ISOGONAL CONJUGATES

In this section and the next, three major contributions to the synthetic geometry of the triangle, all made within the past century, are discussed. The first of these, credited to Lemoine in about 1873, has to do with the concept of symmedians.

Consider Figure 4.21. If $\overline{AE}$ is the bisector of angle A and if $\angle DAE \cong \angle FAE$, then $\overline{AD}$ and $\overline{AF}$ are called *isogonal lines* and one is

FIGURE 4.21

called the *isogonal conjugate* of the other. The bisector of the angle is the bisector of the angle between two isogonal conjugates. Note that, according to this definition, the adjacent sides of the triangle are themselves isogonal conjugates.

DEFINITION. A *symmedian* is the isogonal conjugate of a median.

Suppose, in Figure 4.21, that $\overline{AF}$ is a median. Then $\overline{AD}$ is a *symmedian*. The three symmedians of a triangle have been investigated in great detail by Lemoine and other more recent geometers. They have properties analogous in many ways to the medians themselves. Only a few examples are included in this section. The next theorem includes symmedians as a special case.

THEOREM 4.20. The isogonal conjugates of a set of concurrent segments from the vertex to the opposite sides of a triangle are also concurrent.

Ceva's theorem should suggest itself as a tool to use in proving Theorem 4.20, but the form of the theorem needs to be changed so that ratios of angles are involved.

In Figure 4.22, assume first that G is the point of concurrency of three segments and that $\overline{AD}$, $\overline{BE}$, and $\overline{CF}$ are the isogonal conjugates of these segments.

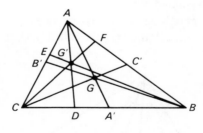

FIGURE 4.22

Consider triangles $AA'C$ and $AA'B$. By the Law of Sines,

$$\frac{CA'}{\sin CAA'} = \frac{CA}{\sin CA'A},$$

so that

$$CA' = \frac{CA \sin CAA'}{\sin CA'A}.$$

Similarly,

$$A'B = \frac{AB \sin BAA'}{\sin BA'A}.$$

Since the sines of the supplementary angles at A' are equal,

$$\frac{CA'}{A'B} = \frac{CA \sin CAA'}{AB \sin BAA'}.$$

Expressing each ratio of division of the sides of triangle ABC in this form leads to a trigonometric form for the theorem of Ceva and its converse:

$$\frac{\sin CAA'}{\sin BAA'} \cdot \frac{\sin BCC'}{\sin ACC'} \cdot \frac{\sin ABB'}{\sin CBB'} = 1.$$

Now in Figure 4.22, because $\overline{AD}$, $\overline{BE}$, and $\overline{CF}$ are isogonal conjugates of the three original segments, various angles can be substituted. For example, $\angle BA'A \cong \angle CAD$ and so on. When all of the appropriate substitutions are made,

$$\frac{\sin BAD}{\sin CAD} \cdot \frac{\sin ACF}{\sin BCF} \cdot \frac{\sin CBE}{\sin ABE} = 1,$$

and $\overline{AD}$, $\overline{BE}$, and $\overline{CF}$ are concurrent by the converse of Ceva's theorem.

If $\overline{AA'}$, $\overline{BB'}$, and $\overline{CC'}$ are medians, then their isogonal conjugates are symmedians; thus a corollary of Theorem 4.20 is:

THEOREM 4.21. The symmedians of a triangle are concurrent.

DEFINITION. The point of concurrency of the symmedians of a triangle is the *symmedian point* or the *Lemoine point*.

It is well known that the median is the bisector of a line

parallel to the base of the triangle opposite the vertex. An analogous property for the symmedian is given in the next theorem.

THEOREM 4.22. The symmedian from one vertex of a triangle bisects any segment antiparallel to the opposite side of the triangle.

In Figure 4.23, let $\overline{AA''}$ be a symmedian and $\overline{DE}$ a segment antiparallel to $\overline{BC}$ so that quadrilateral $DEBC$ can be inscribed in a circle.

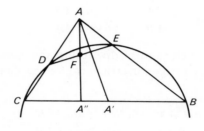

FIGURE 4.23

$$AD \cdot AC = AE \cdot AB,$$

by Theorem 4.9. Also,

$$\frac{\sin \angle CAA'}{\sin \angle A'AB} = \frac{BA}{CA}$$

so that

$$\frac{DF}{FE} = \frac{AD \sin DAF}{EA \sin EAF},$$

$$= \frac{AB}{AC} \cdot \frac{CA}{BA},$$

$$= 1,$$

and F is the midpoint of $\overline{DE}$.

A somewhat different sort of property, one that could be used to describe a symmedian as a set of points satisfying a certain condition, is considered next. The set of points equidistant from the two sides of a triangle lies on a bisector of the angle. There is also a constant ratio of distances from points on a symmedian to the sides.

THEOREM 4.23. The ratio of the distances from a point on the symmedian to the adjacent sides of the triangle equals the ratio of the lengths of those sides.

Use the notation of Figure 4.24, with $\overline{AD}$ the symmedian and $\overline{AA'}$ the median.

FIGURE 4.24

$$\frac{EF}{EG} = \frac{\sin \angle CAD}{\sin \angle BAD}$$

$$= \frac{\sin \angle BAA'}{\sin \angle CAA'} = \frac{AC}{AB}$$

EXERCISES 4.4

1. What is the isogonal conjugate of a symmedian?

2. Prove that the median to a side of a triangle bisects any segment in the triangle parallel to that side.

3. Where is the symmedian point of an equilateral triangle?

4. Could the symmedian point be outside the triangle?

5. Use Theorem 4.23 to answer this question. Assume that the adjacent sides of a triangle are not congruent. Is a point on the symmedian nearer the longer or shorter of the adjacent sides of the angle?

6. Apply Theorem 4.23 to the special case of a right triangle with legs three and four units in length. For each symmedian, state the ratio of the distances from a point on it to the adjacent sides of the triangle.

7. Prove that the ratio of the segments into which a symmedian divides the opposite side of a triangle equals the ratio of the squares of the measures of the adjacent sides.

8. Prove that the altitudes of a triangle are the isogonal conjugates of the circumradii to the vertices of the triangle.

9. Use Theorem 4.20 and Exercise 8 to give an alternative proof that the altitudes of a triangle are concurrent.

4.5   RECENT SYNTHETIC GEOMETRY OF THE TRIANGLE

A second major contribution to the recent geometry of the triangle has to do with the *Brocard points* and the *Brocard circle,* named after Henri Brocard, who proved some theorems about these topics during the last part of the nineteenth century. Like the Miquel point, the Brocard points are defined as special points of intersection of three circles.

THEOREM 4.24. The three circles, each with one side of a triangle as a chord and tangent to the adjacent side, taken in order around the figure, meet in a point called a *Brocard point.*

In Figure 4.25, circle 1 is tangent to $\overline{AC}$, circle 2 is tangent to $\overline{BC}$, and circle 3 is tangent to $\overline{AB}$.

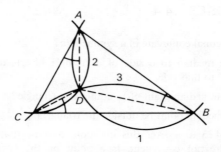

FIGURE 4.25

Assume that circles 1 and 2 meet at *D*. $\angle ABD \cong \angle CAD$ $\cong \angle BCD$, since the angle between a tangent and a chord is, like an inscribed angle, measured by half the intercepted arc. Circle 3 is tangent to $\overline{AB}$ at B, and $\overline{BD}$ subtends $\angle DCB$ equal to $\angle ABD$, so this circle also passes through D. The order chosen was counterclockwise, and point D is called the *first* or *positive Brocard point.*

In general, there are two distinct Brocard points, since the order around the triangle can be either clockwise or counterclockwise. Furthermore, it can be proved that the two Brocard points are isogonal conjugates of each other.

DEFINITION. Points are isogonal conjugates if one is the intersection of lines that are isogonal conjugates of lines intersecting at the other point.

Yet another theorem shows a connection between previous theorems about symmedians and the new theorem about the Brocard points.

THEOREM 4.25. The Brocard points are on the Brocard circle.

DEFINITION. The *Brocard circle* is a circle whose diameter has as endpoints the circumcenter and the symmedian point of a triangle.

In Figure 4.26, let O and S be the circumcenter and the symmedian point, and let the perpendicular bisectors of the sides of the

FIGURE 4.26

triangle meet the Brocard circle again at D', E', and F'. It can be shown that the lines AF', BE', and CD' are concurrent at a Brocard point.

Because right angles are inscribed in a semicircle, $\overline{SD'}$ is parallel to $\overline{BC}$, $\overline{SF'}$ is parallel to $\overline{AC}$, and $\overline{SE'}$ is parallel to $\overline{AB}$. This means that the distances from S to the sides of the original triangle are equal to the distances along the perpendicular bisectors from the points D', E', F' to points D, E, F.

Because of Theorem 4.23,

$$\frac{DD'}{CD} = \frac{EE'}{EB} = \frac{FF'}{FA}$$

so that $\triangle CDD' \sim \triangle BEE' \sim \triangle AFF'$. As a result,

$$\angle BCD' \cong \angle CAF' \cong \angle ABE'.$$

A comparison with Figure 4.25 shows that the three sides of the angles meet at a Brocard point G. Also, since $\angle AF'F \cong \angle GF'O \cong GD'O$, points F', D', O, G all lie on a circle, the Brocard circle, as was to be proved. For an equilateral triangle, the circumcenter and symmedian point coincide so that the Brocard circle reduces to a single point.

The second major topic in this section is a theorem discovered about 1899 by Frank Morley, father of the author Christopher Morley. Its significance is based on the fact that it concerns trisectors of angles rather than bisectors. Angle bisectors of a triangle meet at a point, but it seems that mathematicians have only recently considered what happens to sets of three adjacent trisectors.

THEOREM 4.26. *Morley's theorem.* The adjacent trisectors of the angles of a triangle are concurrent by pairs at the vertices of an equilateral triangle.

Before reading further, you will find it profitable to use a protractor and carefully draw several triangles of various shapes to verify the reasonableness of Morley's theorem.

Since 1900, many different proofs have been given for this theorem, but the one here is an indirect approach, starting with an equilateral triangle and ending with the original given triangle.

Figure 4.27a shows what is meant by adjacent trisectors. For example, $\overline{CD}$ and $\overline{BD}$ are a pair of adjacent trisectors. In Figure 4.27b, assume that $\triangle DEF$ is an equilateral triangle. Isosceles triangles DGE, DHF, and FIE can be determined such that the measures X, Y, Z

of the angles are each less than 60 and $X + Y + Z = 120$. Points A, B, C are determined by extending the sides of these isosceles triangles.

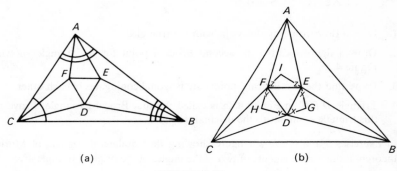

(a)   (b)

FIGURE 4.27

The measures of other angles at points D, E, F can be determined from the given information. According to Exercise 8 of Exercise Set 4.1, the angle formed by the lines containing the segments from an incenter to two vertices is 90 plus one-half the measure of the angle at the third vertex. In triangle CIB,

$$m\angle CIB = 180 - 2Z,$$

whereas $m\angle CDB = 360 - (X + Y + 2Z + 60)$.

$$360 - (X + Y + 2Z + 60) = 360 - (X + Y + Z) - Z - 60$$
$$= 360 - 120 - Z - 60$$
$$= 180 - Z$$
$$= 90 + \tfrac{1}{2}(180 - 2Z)$$

From this discussion, along with Exercise 8 of Exercise Set 4.1, point D is the incenter of $\triangle CIB$. Similarly, F is the incenter of $\triangle AGC$ and E is the incenter of $\triangle AHB$. Thus, the three angles at C are congruent as are the three at A and at B. These angles have measures

$$60 - Y, \quad 60 - Z, \quad 60 - X.$$

The base angles of the isosceles triangle can be found in terms of the angles A, B, C of the original triangle; for example, $x = 60 - \frac{1}{3} \angle B$ so that a triangle ABC can always be found similar to any given triangle.

EXERCISES   4.5

1. Does a Brocard point always lie within a triangle?

2. Draw a figure locating the second Brocard point for the triangle shown in Figure 4.25.

3. Prove that the two Brocard points are isogonal conjugates of each other.

4. Triangle $D'E'F'$ in Figure 4.26 is called the first Brocard triangle. Prove that this triangle is similar to the original triangle.

In Exercises 5 and 6, draw a figure showing the equilateral triangle of Morley's theorem if the measurements of two of the angles of the original triangle are:

5. 90° and 40°.

6. 60° and 20°.

7. Prove that, in Figure 4.27b, the segments $\overline{DI}$, $\overline{FG}$, and $\overline{EH}$ are concurrent.

8. Investigate the figure formed by the intersections of the four pairs of adjacent trisectors of the angles of a square.

9. Investigate the figure formed by the intersections of the four pairs of adjacent four-sectors (or quadrisectors) of the angles of a square, if four-sectors are defined to be the rays partitioning the angle and its interior into four congruent angles and their interiors.

4.6   SPECIAL APPLICATIONS OF EUCLIDEAN GEOMETRY

Applications of synthetic Euclidean geometry to be discussed in this section include the golden ratio, tessellations, and caroms. Each of these is of current interest to mathematicians.

With their sense of beauty and proportion, the Greeks came to regard certain shapes as more pleasing than others and to build many of their buildings in these shapes. The most famous example of this is what is called the *golden ratio,* the ratio of the lengths of the sides of a rectangle with the most pleasing proportion.

In Figure 4.28, suppose that $AB/BC = AC/AB$. In this case, point B is said to divide AC in *extreme and mean ratio,* and, if the length of $\overline{AB}$ is one, the length of $\overline{AC}$ is the *golden ratio.*

FIGURE 4.28

Let x represent the numerical value of the golden ratio. From the previous proportion,

$$\frac{1}{x-1} = \frac{x}{1},$$

$$1 = x^2 - x,$$

$$x^2 - x - 1 = 0,$$

$$x = \frac{1 \pm \sqrt{5}}{2}.$$

The positive value of x,

$$\frac{1 + \sqrt{5}}{2},$$

is the numerical value of the golden ratio. This, of course, is an irrational number. An approximate decimal value is 1.62. Mathematicians have also shown that the ratios of the $(n + 1)$ to the nth term in the set of Fibonacci numbers 1, 1, 2, 3, 5, 8, 13, 21, ... n approach the golden ratio as a limit. The reader should express 3/2, 5/3, 8/5, and so on as decimals to see that this statement is reasonable.

In Figure 4.29, $ABCD$ is a golden rectangle, with $AB = x$ the

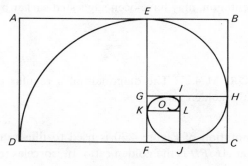

FIGURE 4.29

golden ratio and $\overline{BC}$ the unit segment. If square $AEFD$ is removed, the remaining rectangle $EBCF$ is also a golden rectangle.

It is necessary to show that $CB/EB = x$. Since $CB = 1$ and $BH = x - 1$,

$$\frac{1}{x - 1} = \frac{CB}{BH}.$$

Or, since $BH = EB$, the proportion may be written as

$$\frac{CB}{EB} = \frac{1}{x - 1}.$$

From the equation $x^2 - x - 1 = 0$,

$$1 = x^2 - x,$$

$$\frac{1}{x} = x - 1.$$

Then

$$\frac{CB}{EB} = \frac{1}{1/x},$$

$$= x.$$

This process of removing squares may be continued, and each time the result is a smaller golden rectangle. As shown in the figure, the pattern suggests a spiral, called the *golden spiral*, whose equation in polar coordinates is $r = x^{2\theta/\pi}$, where x is the golden ratio.

One additional application of the golden ratio, its connection to a regular pentagon, may have been suggested earlier by the expression

$$x = \frac{1 \pm \sqrt{5}}{2}.$$

THEOREM 4.27. The diagonals of a regular pentagon divide each other in the golden ratio.

The notation of Figure 4.30 is used to illustrate Theorem 4.27 by showing that AF/FD is the golden ratio. In isosceles trapezoid $ACDE$,

$\triangle ACF \sim \triangle DFE$, so $AF/FD = AC/ED$. But $ED \cong BC$ and $\overline{BC} \cong \overline{AF}$, since $ABCF$ is a rhombus. Thus,

$$\frac{AF}{FD} = \frac{AC}{AF}, \quad \text{or} \quad \frac{AF}{FD} = \frac{AD}{AF},$$

so AF/FD is the golden ratio. Note that if the sides of a regular pentagon are one unit, then the length of the diagonals is the numerical value of the golden ratio.

FIGURE 4.30

The second application of synthetic Euclidean geometry to be discussed is that of tessellations.

DEFINITION. A *tessellation* is a pattern of polygons fitted together to cover an entire plane without overlapping. *Regular tessellations* are those in which all of the polygons are congruent and regular, with common vertices.

Figure 4.31 shows regular tessellations with squares, equilateral triangles, and regular hexagons. The fact that these are the only three regular tessellations is established in the proof of the following theorem.

(a) (b) (c)

FIGURE 4.31

THEOREM 4.28. The only three regular tessellations of the plane are those with the square, the equilateral triangle, and the regular hexagon.

Let x be the number of sides for the polygon and y the number of polygons coming together at each vertex. Then the two expressions for the measure of the angle can be set equal.

$$\frac{\pi(x - 2)}{x} = \frac{2\pi}{y}$$

$$\pi xy - 2\pi y = 2\pi x$$

$$xy = 2x + 2y$$

$$xy - 2x - 2y = 0$$

$$xy - 2x - 2y + 4 = 4 \qquad \text{(add 4 to both members)}$$

$$(x - 2)(y - 2) = 4$$

There are only three ways to factor 4, and these three ways lead to the three tessellations previously named.

TABLE 4.1

$x - 2$	$y - 2$	x	y	Tessellation
4	1	6	3	Figure 4.31c
2	2	4	4	Figure 4.31a
1	4	3	6	Figure 4.31b

Sometimes, the definition of tessellation is broadened to include figures other than polygons. The art of M. C. Escher, mentioned in Chapter 2, provides examples of somewhat more general plane-filling drawings.

One of the ancient yet very modern applications of the Euclidean geometry of similar triangles is today called *caroms,* after the word carom, which means an impact followed by a rebound. Historically, applications of this sort are based on an extremum property stated as Heron's theorem.

THEOREM 4.29. *Heron's theorem.* For two points on the same side of a line, the shortest path from the first point to the line and then to the second point is by way of the point of intersection of the line and the segment from the first point to the reflection of the second point.

In Figure 4.32, we need to prove that $AC + CB$ is a minimum. This means that $AC + CB < AC' + C'B$ for any other point C' on the given line.

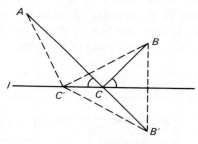

FIGURE 4.32

$CB = CB'$ and $BC' = B'C'$, for B' the reflection of B. Then

$$AC + CB = AC + CB' = AB',$$

and

$$AC' + C'B = AC' + B'C'.$$

But $AB' < AC' + B'C'$, since $AC'B'$ is a triangle, so $AC + CB < AC' + C'B$, as was to be established.

The physical applications of Heron's theorem depend on the fact that the angles $\overline{AC}$ and $\overline{CB}$ make with line l are congruent. For example, the shortest path from A to B by way of a point on line l is the path of a light ray reflected in a mirror at line l.

The problem can be generalized for more than one line, thus increasing the usefulness of the theory. For example, Figure 4.33 shows the shortest path from A to B, touching l_1 and then l_2. B' is the reflection of B in l_2 and B'' is the reflection of B' in l_1. The path leads from A towards B'', then from C towards B'.

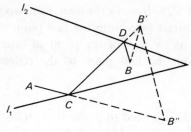

FIGURE 4.33

An application of the theory of caroms that is of current interest in mathematics as well as in everyday life is in connection with a pool table. Figure 4.34 shows how the ball at A can be aimed to hit ball B by caroming off one, two, or three sides of the table.

(a)   (b)   (c)

FIGURE 4.34

Heron's theorem and the transformation of reflection also provide the basis for other applications of geometry to *extremum problems* (those involving maxima and minima). Two examples, solved without the use of calculus, are included.

THEOREM 4.30. The triangle with a given area and side, for which the sum of the measures of the other two sides is a minimum, is an isosceles triangle.

The proof is outlined, but the details are left as Exercise 12, Exercise Set 4.6. In Figure 4.35, if $\overline{AB}$ is the fixed side, then the third vertex X must be on a line $\overleftrightarrow{CD}$ parallel to $\overleftrightarrow{AB}$. Why? By Heron's theorem, $AX + XB$ is a minimum if $\triangle AXB$ is isosceles. Why?

FIGURE 4.35

A second example, called *Fagnano's problem* after J. F. Fagnano, who proposed it in 1775, is stated as the last theorem of the chapter.

THEOREM 4.31. In a given acute-angled triangle, the inscribed triangle with minimum perimeter is the orthic triangle, whose vertices are the feet of the altitudes.

Details of the proof are again left as an exercise (Exercise 13, Exercise Set 4.6). In Figure 4.36, if $\triangle DEF$ is an arbitrary inscribed

FIGURE 4.36

triangle and if G and H are the images of D under reflection with respect to $\overleftrightarrow{AC}$ and $\overleftrightarrow{AB}$, then the distance $GE + EF + FH$ is the perimeter of the inscribed triangle. Why? This distance will be a minimum if G, E, F, H are collinear. The measure of $\angle GAH = 2m\angle CAB$. Why? $\triangle GAH$ is isosceles. Why? Because the measure of $\angle GAH$ is independent of the choice of point D, the base $\overline{GH}$ will have a minimum measure when the congruent sides are a minimum. But this condition is met when $\overline{AD}$ has a minimum length, and this occurs when $\overline{AD}$ is an altitude. Why?

EXERCISES 4.6

1. Find the approximate value of the golden ratio correct to four decimal places.

2. Use Fibonacci numbers to give a series of approximations to the golden ratio until one is reached that is correct to three places past the decimal point.

3. In Figure 4.29, if $BC = 1$, find the measure of segment $\overline{IL}$.

4. In Figure 4.30, prove that $ABCF$ is a rhombus.

5. Draw a tessellation that is not a regular tessellation.

6. Show by a drawing and a numerical explanation why regular octagons cannot be used for a regular tessellation.

7. Use Heron's theorem to prove that, given the length of one side and the measure of area, the triangle for which the sum of the other two sides is smallest is isosceles.

Draw figures similar to Figure 4.37 for the following circumstances.

FIGURE 4.37

8. Where should ball A hit side $\overline{CD}$ in Figure 4.37a in order to carom and hit ball B?

9. Ball A in Figure 4.37b is to carom off $\overline{DE}$, then $\overline{EF}$, and then hit ball B.

10. Ball A in Figure 4.37c is to carom off side $\overline{CD}$, then $\overline{DE}$, then $\overline{EF}$, and finally hit ball B.

11. Using the notation of Figure 4.33, prove that the shortest path from A to B, by way of l_1 and then l_2, is the path shown.

12. Complete the proof of Theorem 4.30.

13. Complete the proof of Theorem 4.31.

CHAPTER 5

CONSTRUCTIONS

5.1 THE PHILOSOPHY OF CONSTRUCTIONS

The concept of constructions in geometry can be traced back at least as far as the work of Plato, the great Greek philosopher who lived about a century before the time of Euclid. It was Plato who is given credit for specifying the use of the *straightedge* and the *compass* as the only two instruments for performing constructions. Technically, Plato specified the use of *dividers,* or a *collapsing compass,* not the modern compass we are accustomed to. When the instrument had been used to construct one circle, the measurement was lost because the instrument could not be moved to construct another circle with the same radius. This characteristic and the limitations of the unmarked straight edge account for the fact that the transfer of distance was not postulated by the Greeks. It will be shown, however, that the divider (collapsing compass) is equivalent to the modern compass, so the straightedge and the compass are still the only two instruments allowed in the classical mathematical constructions of Euclidean geometry. Constructions that are almost trivial with a modern compass become more complicated with the original compass. Nevertheless, the reader should check the great variety of construction problems specified in the books of Euclid, all of which are possible with the unmarked straightedge and the collapsing compass.

The concept of *mathematical construction* is difficult to explain because the word "construction" is used in at least three ways:

1. To describe the geometric problem to be solved.
2. To describe the process of solving the problem.
3. To describe the completed drawing that results from solving the problem.

The result of a construction is a drawing that shows certain relationships among lines and circles. Philosophically, constructions may be explained as methods for solving certain geometrical problems according to a fixed set of rules. This will become clearer when other basic ideas have been presented.

The concept of a construction was basic to the axiomatic system of Euclid, as can be seen by reviewing his axioms and postulates stated in Chapter 1. On the other hand, modern postulates for Euclidean geometry represented by the sets in Appendices 1–3 make no mention of constructions, and it may be stated that constructions are outside the strict axiomatic development of modern Euclidean geometry. Yet the concept of a construction remains significant in geometric thought. Constructions are studied in various modern geometries not only because they are interesting in their own right but also because they provide applications of other geometric concepts.

The problem in a construction is not simply that of drawing a figure to satisfy certain conditions but whether, by using a compass and straightedge only, a theoretically exact solution can be obtained. The drawings in construction problems are approximations, but the theory is exact. The first three postulates of Euclid provide the axiomatic basis for the constructions of the Greeks. The first two of these postulates make it possible to construct any portion of a straight line through two points, whereas the third makes it possible to construct a circle if the center and the measure of the radius are given.

An example of a simple construction problem solved with the modern compass and then with the collapsing compass will illustrate the difference in the methods that must be used. Figure 5.1a shows the familiar method of finding the midpoint of a segment by construction.

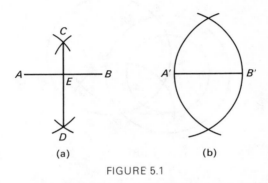

(a)   (b)

FIGURE 5.1

Note that $\overline{AC}$ and $\overline{BC}$ must be congruent, that $\overline{AD}$ and $\overline{BD}$ must also be congruent, but that $\overline{AC}$ and $\overline{AD}$ are not necessarily congruent. Figure 5.1b shows the same problem solved with a collapsing compass. In this case, the measure of $\overline{A'B'}$ is used for the radius since its length can be determined using either endpoint as the center of the arc, whereas the arbitrary radius AC in Figure 5.1a cannot be reproduced again with the second endpoint B as the center of the circle.

THEOREM 5.1. The compass and the collapsing compass (dividers) are mathematically equivalent.

The proof of this theorem consists of showing that a circle can be constructed with the collapsing compass, given the center and two other points that determine the length of the radius. The steps in performing this construction are stated, using the notation of Figure 5.2. The problem is to construct a circle with center A and with radius BC.

1. Construct the circle with center A and passing through B.
2. Construct the circle with center B and passing through A. These two circles meet at D and E.
3. Construct the circle with center E and passing through C.
4. Construct the circle with center D and passing through C.
5. The circles in steps 3 and 4 intersect again in a point F, and the circle with center A and radius AF is the required circle.

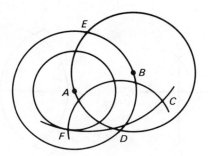

FIGURE 5.2

The proof, using congruent triangles, that $\overline{AF}$ is congruent to $\overline{BC}$ is left as an exercise.

It is assumed that you are familiar with the basic constructions of Euclidean geometry. These include transferring a segment, bisecting a segment, constructing a perpendicular to a line at a certain point, constructing an angle bisector, copying an angle, constructing a triangle given an angle and the two adjacent sides, constructing a triangle given the three sides, and constructing a line through a point parallel to a given line. It is also quite possible that you have encountered some of the more difficult constructions to be introduced later in the chapter.

Several additional basic constructions are introduced or reviewed here so that they may be used easily.

Partition of a segment into n congruent segments, for n a positive integer greater than one, is the first of these constructions and is illustrated in Figure 5.3, for $n = 5$. The problem is to partition $\overline{AB}$ into five congruent segments. $\overline{AC}$ is constructed through A at any convenient angle. Then an arbitrary unit AD is chosen to determine five congruent segments along $\overleftrightarrow{AC}$. Next $\overline{HB}$ is constructed, and the four segments $\overline{DI}$, $\overline{EJ}$, $\overline{FK}$, and $\overline{GL}$ are constructed parallel to $\overline{HB}$, through the established points on $\overleftrightarrow{AC}$, determining the four required points I, J, K, L.

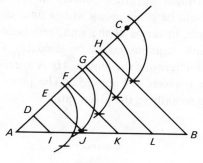

FIGURE 5.3

A second problem related to the first is to *partition a segment into a given ratio.* The problem is to find a point B on $\overleftrightarrow{AC}$ such that AB/BC has a given numerical value. One use of this construction is in the theory of harmonic sets of points in projective geometry. Figure 5.4

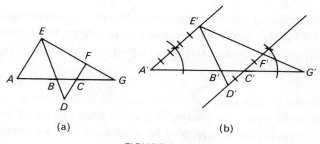

(a)   (b)

FIGURE 5.4

is an *analysis figure,* showing the completed solutions. It is assumed that B and G divide $\overline{AB}$ internally and externally in the desired ratio. If $\overline{AE}$ and $\overline{DF}$ are parallel, then $\triangle AEB \sim \triangle CDB$ and $\triangle AEG \sim \triangle CFG$, and the ratios of similarity for the pairs of similar triangles are the given ratio of division. Then the given ratio is also the ratio of AE to CD and of AE to CF; this makes possible the following construction, illustrated in Figure 5.4b for a ratio of 5/2. Through A' and C', construct parallel lines. Using an arbitrary unit, find E' so that $A'E' = 5$ and find

D' and F' so that $C'D' = C'F' = 2$. Connect E', D' and E', F' to locate the required points B' and G'.

In the problem of partitioning a segment into a given ratio, the known information might be given entirely in the form of segments. For example, $\overline{AC}$ could be given along with a unit segment and a third segment whose length, in terms of the unit, represents the given ratio. In this case, the unit segment could be used for $C'D'$, whereas the third segment would determine the length AB. A specific example of this type of construction is given in Figure 5.5. The problem is to partition $\overline{AC}$ internally and externally in the given ratio.

FIGURE 5.5

The use of the basic constructions of this section leads to the solution of more complicated construction problems. In a formal study of constructions, four distinct steps are required in the solution of any construction problem.

1. *Analysis.* In this step, the solver assumes that the construction has been performed, then analyzes the completed picture of the solution to find the needed connections between the unknown elements in the figure and the given facts in the original problem.
2. *Construction.* The result of this step is the drawing itself, made with straightedge and compass and showing the construction marks.
3. *Proof.* It is necessary to prove that the figure constructed is actually the required figure.
4. *Discussion.* The number of possible solutions and the conditions for any possible solution are explained in this step.

In this text, it will not be necessary to carry through all four steps in complete detail, although each will be illustrated.

EXERCISES 5.1

For Exercises 1 and 2, explain how a collapsing compass can be used to:
1. Bisect an angle.
2. Transfer a segment.
3. Complete the proof of Theorem 5.1.

In Exercises 4–6, show how to perform these basic constructions in Euclidean geometry:
4. Construct a perpendicular to a line at a certain point.
5. Construct an angle congruent to a given angle.
6. Construct a line through a point and parallel to a given line.

For Exercises 7–10, carry out the construction indicated.
7. Partition a given segment into seven congruent segments.
8. Partition a given segment internally and externally in the ratio of three to two, given a unit segment.
9. Carry out the same construction as in Exercise 8, but use a ratio of three to four.
10. Partition a given segment internally and externally in the ratio of length of two given arbitrary segments, neither of which is the unit segment.

5.2   CONSTRUCTIBLE NUMBERS

The weaknesses of Greek algebra and the strengths of Greek geometry were all too apparent. Without an adequate notation, they used geometric constructions to solve many of the problems now solved algebraically—for example, the solution of some algebraic equations. But one thing that this did accomplish was to focus attention on the connections between geometry and the various number systems of algebra, a study that has continued to have significance in both modern algebra and modern geometry.

A unit segment represents the number one. What other numbers can also be represented by segments, beginning with this unit segment and using only the straightedge and compass to construct other segments? The answer to this question defines what is known as the set of *constructible numbers*. Figure 5.6 shows a geometric interpretation of the four rational operations on whole numbers, as well as the construction of some irrational numbers by the process of extracting the square root of a positive rational number. The given information consists of the three segments in Figure 5.6a.

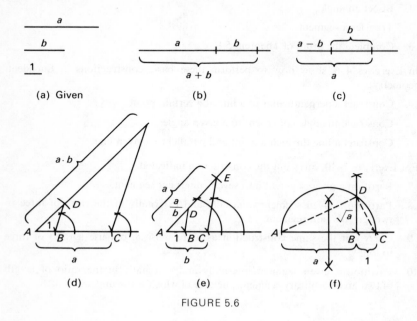

(a) Given   (b)   (c)

(d)   (e)   (f)

FIGURE 5.6

The diagrams in Figure 5.6b and c for addition and subtraction should be self-explanatory. For multiplication, the proof of the construction in Figure 5.6d depends on the proportion

$$\frac{1}{b} = \frac{a}{AE}$$

so that $AE = ab$.

For division, the proof of the construction in Figure 5.6e also depends on a proportion,

$$\frac{1}{AD} = \frac{b}{a}, \quad \text{or} \quad AD = \frac{a}{b}.$$

The proof that BD is equal to $\sqrt{a}$ in Figure 5.6f likewise depends on a proportion that in turn is derived from similar right triangles in the figure.

$$\triangle ADB \sim \triangle DCB,$$

so that

$$\frac{AB}{BD} = \frac{BD}{BC},$$

$$AB = (BD)^2,$$

$$\sqrt{AB} = BD.$$

The first four constructions of addition, subtraction, multiplication, and division make it possible to construct a segment representing any number in the field of rational numbers, given the unit segment. The construction of the square root makes it possible to construct numbers in *extension fields* having the field of rational numbers as a subset. Recall that a field of numbers has the closure property for rational operations, with division by zero excluded. Examples of extension fields of constructible numbers are $a + b\sqrt{2}$, $a + b\sqrt{3}$, $\sqrt{a + b\sqrt{2}}$, for a and b rational numbers and the entire radicand positive in the third example. For example, the number $5 + \sqrt{3 + \sqrt{7 + \sqrt{2}}}$ is a constructible number, while $\sqrt[3]{7}$ is not.

A somewhat more general approach to constructible numbers can be studied from an algebraic viewpoint. Suppose that all of the numbers in some number field F can be constructed.

THEOREM 5.2. The use of a straightedge alone can never yield segments for numbers outside the original number field.

The equations for any two pairs of lines through distinct pairs of points with coordinates (a, b), (c, d), (e, f), (g, h) in a field are

$$y - b = \frac{d - b}{c - a}(x - a)$$

and

$$y - f = \frac{h - f}{g - e}(x - e).$$

The point of intersection of these two lines has coordinates obtained by rational operations on elements of the field F, hence the use of the straightedge alone did not result in a number outside the original field. It is naturally assumed that division by zero is avoided.

Now select an element a of F such that $\sqrt{a}$ is not an element of F. All numbers of the type $b + c\sqrt{a}$ (b and c also in F) can be constructed with the use of the straightedge and compass. Numbers of the form $b + c\sqrt{a}$ themselves constitute a field, and this is an extension field of F. A single use of the compass cannot lead from F beyond an extension field of F, however.

THEOREM 5.3. A single application of the compass using numbers of a number field results only in elements of the extension field $b + c\sqrt{a}$, where a, b, and c are elements of the original field, with a positive.

From an algebraic point of view, it is necessary to consider the intersection of a circle and a straight line and then the intersection of two circles in order to prove the theorem.

The intersection of a circle, $x^2 + y^2 + ax + by + c = 0$, and a straight line, $dx + ey + f = 0$, with all coefficients in F, is given by the solutions of

$$x^2 + \left(\frac{-f - dx}{e}\right)^2 + ax + b\left(\frac{-f - dx}{e}\right) + c = 0.$$

This equation may be written in the form

$$\left(1 - \left[\frac{d}{e}\right]^2\right)x^2 + \left(\frac{2df - bd}{e} + a\right)x + \left(\frac{f^2 - bf}{e} + c\right) = 0.$$

Each of the coefficients of this equation is an element of F, so they may be indicated as g, h, k, so that $gx^2 + hx + k = 0$. The quadratic formula yields the solutions

$$x = \frac{-h \pm \sqrt{h^2 - 4gk}}{2g}.$$

Both of these solutions are of the form $b + c\sqrt{a}$, for certain numbers in F, so that the use of the compass did not lead outside the extension field. The proof of the second part of the theorem involving the intersection of two circles is left as an exercise. As a result of this analysis, constructible numbers can be characterized as those that can be obtained through a sequence of extension fields of the type discussed.

EXERCISES 5.2

Take a given unit segment and two other given segments to perform the constructions in Exercises 1–5.

1. Find a segment whose measure is the sum of the measures of the two given segments.

2. Find a segment whose measure is the difference of the two given segments.

3. Find a segment whose measure is the product of the measures of the two given segments.

4. Find a segment whose measure is the quotient of the measures of the two given segments.

5. Find a segment whose measure is the square root of the measure of the longer segment.

For Exercises 6–10, with a given unit segment, construct segments for each of these numbers in the set of constructible numbers. Use the results of one exercise for the next.

6. 3   7. $\sqrt{3}$ 8. $2 + \sqrt{3}$ 9. $\sqrt{2 + \sqrt{3}}$ 10. $\sqrt{3 + \sqrt{2 + \sqrt{3}}}$

11. Construct a segment for $\sqrt{2/3}$, given a unit segment.

12. In Theorem 5.2, find the point of intersection of the two lines.

13. In the proof of Theorem 5.2, show that numbers of the form $b + c\sqrt{a}$ are closed under the four rational operations.

14. Complete the proof of Theorem 5.3 for two intersecting circles.

5.3 CONSTRUCTIONS IN ADVANCED EUCLIDEAN GEOMETRY

In traditional college geometry, the construction of triangles played a dominant role. One reason for this was that constructions gave a ready application of many of the new concepts introduced. In some modern intuitive geometry courses, especially at the beginning secondary level, and in college courses for elementary teachers of mathematics, constructions serve as a primary vehicle to present concepts for the first time.

The reader of this section should get some idea of the variety of construction problems and should also appreciate the use of constructions in the application of concepts already studied in previous chapters. Developing great skill in using constructions is not the desired outcome. Often only a brief analysis of the problem will be presented, rather than the detailed construction itself. The proof and the discussion will sometimes be suggested.

Construction problem. Construct a triangle, given the length of one side of the triangle and the lengths of the altitude and the median to that side.

The analysis figure is shown in Figure 5.7a. From the given information, right triangle $AA'D$ can be constructed immediately since two sides are known. Then points B and C can be located on $\overleftrightarrow{DA}$. Each is a distance half the measure of the given side from the determined midpoint A'. The actual construction is shown in Figure 5.7b. Triangle $AA'D$, and hence the required triangle, can always be constructed as long as the segment for the median is at least as long as the segment for the altitude. There is only one possible solution. Triangle $AA'D$ is an example of an *auxiliary triangle* in a construction. An auxiliary triangle, usually a right triangle, is one that can be constructed immediately from the given information.

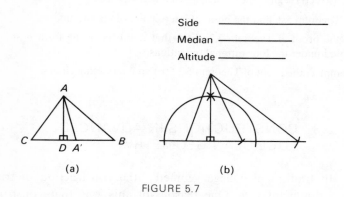

FIGURE 5.7

The second example illustrates the important idea of locating one of the vertices of the required triangle as the intersection of two sets of points satisfying given conditions. Traditionally, this point is described as the intersection of two *loci*, a word meaning "paths."

Construct a triangle, given one angle, the length of the side opposite this angle, and the length of the altitude to that side. An analysis figure is shown in Figure 5.8a. Assume that $\angle BAC$, AD, and BC are given. Since B and C can be located from the given information, the only remaining problem is to locate point A relative to B and C. The other conditions do not determine A individually, but they can be used together. One condition for the location of A is that it must be on a line parallel to $\overline{BC}$ and at a distance of AD from the opposite side. Vertex A must also lie on the circumcircle, and that can be found from knowing the angle and the opposite side. This *auxiliary construction* is shown in Figure 5.8b. The two conditions determine A, and the triangle can be constructed.

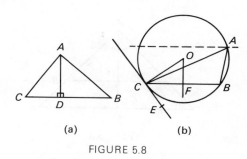

(a)   (b)

FIGURE 5.8

The auxiliary problem in this example is interesting in its own right. In Figure 5.8b, the angle FCE is the given angle, so the center of the circle can be located at the intersection of the perpendicular bisector of $\overline{BC}$ and the perpendicular to $\overline{CE}$ at C.

The three elements consisting of the measure of one angle of a triangle, the length of the opposite side, and the radius of the circumcircle constitute an example of a *datum*.

DEFINITION. A datum is a set of n elements, any $n - 1$ of which determine the remaining one.

Figure 5.8b shows only one part of the proof that the three elements constitute a datum, since it must be shown that *each two*

determine the third. The concept of a datum is important in constructions, since two given elements will determine a third element if the
three constitute a datum.

The discussion of the problem represented by Figure 5.8 consists
of determining the number of solutions. There could be two lines
parallel to $\overline{BC}$, and each of these might intersect the circle twice. In
one case, however, the angle would be the supplement of the given
angle, so that case must be discarded. This leaves the possibility for
two, one, or no solutions, depending on how many times the line intersects
the circumcircle.

Often, the concept of similar triangles can be used in constructions. This is illustrated in the problem of *constructing a triangle,
given the measure of two angles and the length of the bisector of the
third angle.* Figure 5.9 is the analysis figure. The measures of the three

FIGURE 5.9

angles of a triangle constitute a datum, since any two determine the
third. From this information, a triangle $AB'C'$, similar to the required
triangle, can be constructed. This idea is known as the triangle being
determined in *species.* A family of triangles is determined, and the required
triangle can be found as one particular member of this family. The
required triangle is found by laying off the given length of the bisector
along $\overline{AD'}$ to locate point D and then constructing the parallel to
$\overline{C'B'}$ through D. There is always one solution as long as the sum of the
measures of the two given angles is less than π.

Construction can involve concepts from Euclidean geometry that
are more advanced than those used so far in this section. One
example of this is to *construct a triangle, given the circumcenter, the
center of the nine-point circle, and the midpoint of one side.*

The analysis figure is shown in Figure 5.10. The location of

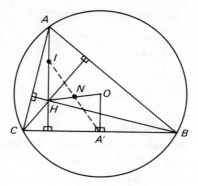

FIGURE 5.10

H, the orthocenter, is determined from the fact that the nine-point center, N, is the midpoint of the segment joining the orthocenter and the circumcenter O. Since N is also the midpoint of $\overline{A'I}$, where I is the midpoint of $\overline{HA}$ and A' is the midpoint of the side, point I can be determined, then vertex A can be found a known distance on $\overleftrightarrow{HI}$ beyond I. One locus for points B and C is a line through A' perpendicular to $\overline{OA'}$. The second locus is the circumcircle, and this can be determined because the radius OA has been found.

A second example of a construction using a more advanced concept is the *construction of a triangle, given the lengths of the altitude, median, and symmedian from the same vertex*. The analysis figure is shown in Figure 5.11. $\overline{AD}$ is the altitude, $\overline{AE}$ is the symmedian, and $\overline{AA'}$ is the median. The analysis and discussion are left as an exercise.

FIGURE 5.11

Construction problems do not always involve constructing a triangle as the completed figure. For example, suppose it is required to *construct a rectangle, given the lengths of one side and one diagonal.*

FIGURE 5.12

The analysis figure, Figure 5.12, shows that each of the two right triangles *BCD* and *ABD* can be constructed from the given information.

A final example is that of *constructing a circle with a known radius that is tangent to a given line and orthogonal to a given circle.* Two curves are *orthogonal* if they meet at right angles. The analysis figure is shown in Figure 5.13. The circle with center *O* is the required circle, *l* is the given tangent, and the circle with center *O'* is the given circle.

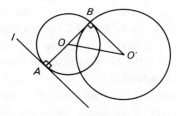

FIGURE 5.13

The only thing to be determined is the location of the center of the required circle. One condition is that the center lie on a line parallel to *l* at a given perpendicular distance *AO* from line *l*. The second condition is that the center lie a known distance of $OO' = \sqrt{(OB)^2 + (BO')^2}$ from *O'* so that its position can be fixed. Because the location of *O* depends on the intersection of a circle with two parallel lines, a complete discussion of the problem would have to consider from four down to zero possible solutions.

EXERCISES   5.3

1. Finish the proof that the measure of one angle of a triangle, the measure of the opposite side, and the radius of the circumcircle constitute a datum.

2. Draw the actual construction figure corresponding to Figure 5.10.

3. Complete the discussion of the number of solutions in Figure 5.13.

For each of the following constructions:
  a. Sketch an analysis figure.
  b. Give the analysis.
  c. Briefly discuss the number of possible solutions.

4. Construct a triangle, given the measure of one angle, the length of an adjacent side, and the length of the altitude to that side.

5. Construct a triangle, given the measure of one angle, the length of that internal angle bisector, and the length of one adjacent side.

6. Construct a triangle, given the length of one side and the lengths of the medians to the other two sides.

7. Construct a triangle, given the length of one side, the length of the altitude to a second side, and the circumradius.

8. Construct a triangle, given the length of one side, the length of the median to that side, and the circumradius.

9. Construct a circle with a given radius and tangent to two given intersecting lines.

10. Construct a circle with a given radius tangent to a given line and tangent to a given circle.

11. Construct a triangle given the length of one side, the length of the median to that side, and the ratio of the lengths of the two remaining sides.

12. Construct a triangle, given the measure of one angle, the measure of the opposite side, and the radius of the incircle.

13. Construct a triangle, given the measure of one angle, the length of the internal bisector of that angle, and the radius of the incircle.

14. Construct a triangle, given the measure of one angle and the length of the altitudes to the two adjacent sides.

15. Construct a triangle, given the length of one side, the length of the median to that side, and the length of one other median.

16. Construct a triangle, given the measure of two angles and the length of the median to the third side.

17. Construct a triangle, given the measure of one angle, the length of the altitude to the opposite side, and the ratio of the two adjacent sides.

18. Construct a triangle, given the orthocenter, the nine-point center, and the foot of one altitude.

19. Construct a triangle, given the lengths of the altitude, median, and symmedian from one vertex.

5.4 CONSTRUCTIONS AND IMPOSSIBILITY PROOFS

While the Greeks were deeply interested in the philosophy of constructions using the instruments specified by Plato, they also realized that some problems were beyond their abilities to solve by these means. Rather than give up, they invented new instruments that would make the constructions possible. However, not until the nineteenth century did advances in the algebra of the real-number system make it possible to identify constructions that were impossible with the straightedge and compass.

In the history of mathematics, three construction problems became so famous that they are called the "Three Famous Greek Problems." These are the problems of *doubling the cube,* of *trisecting the angle,* and of *squaring the circle.* The algebra of the constructible numbers has provided the information necessary to prove that each construction is impossible.

Various legendary stories account for the origin of the problem of doubling the cube. One states that a king wanted to double the size of the cubical tomb of his son. Another tells that the Delians were instructed by their oracle to double the size of the altar erected to Apollo in order to rid the city of a plague. In these problems, it is the volume of the cube, not the edge, that is to be doubled (see Figure 5.14).

FIGURE 5.14

If the original edge is one unit and the required edge is x units, then the original volume is one cubic unit and the volume of the doubled cube is two units. This means that x must be a real solution of the equation $x^3 = 2$. It was established in Section 5.2 that the set of constructible numbers must be an element of some extension field of the set of rational numbers. The assumption that the solution of $x^3 = 2$

lies in one of these extension fields leads to a contradiction. It can be proved, although the algebraic details are omitted here, that if a number of the form $a + b\sqrt{c}$ is a solution of a quadratic equation, then so is $a - b\sqrt{c}$. It has previously been shown that elements of extension fields can always be written in the form $a + b\sqrt{c}$. But the assumption of two real roots for $x^3 = 2$ contradicts the known fact that two of the cube roots of 2 are nonreal. Thus, x is not a constructible number.

This completes the proof of the impossibility of solving the first of the Greek problems.

THEOREM 5.4. The construction of doubling the volume of a cube cannot be performed by straightedge and compass alone.

The second problem, that of trisecting a general angle, can be disposed of in a way similar to the first. Ordinarily, this problem is approached by showing that a single example, such as trisecting an angle of 60 degrees, is impossible, and hence the general case is

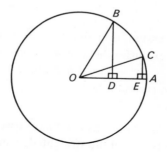

FIGURE 5.15

impossible. In Figure 5.15, suppose that $m\angle BOA = 60$ and $m\angle COA = 20$. If the circle is a unit circle, then $OD = \cos\angle BOA$ and $OE = \cos\angle COA$. The problem of constructing the smaller angle is equivalent to that of finding OE, given OD. The trigonometric identity relating the cosines of an angle and a second angle with one-third the measure of the first angle is

$$\cos\theta = 4\cos^3\left(\frac{\theta}{3}\right) - 3\cos\left(\frac{\theta}{3}\right).$$

If $\cos(\theta/3) = x$, and $\cos\theta$ is a given constant $\frac{1}{2}$ (since $\theta = 60°$), then

$$4x^3 - 3x - \tfrac{1}{2} = 0,$$

or

$$8x^3 - 6x - 1 = 0.$$

This equation has no rational solutions, because if a/b is a rational solution, then a is a positive or negative factor of 1 and b is a positive or negative factor of 8; all possibilities can be checked quickly by synthetic division to see that none is a solution.

If the equation has a solution of the form $a + b\sqrt{c}$, it has $a - b\sqrt{c}$ as another solution. Suppose these represent the least-inclusive extension field of which the solution is a member. The sum of the three roots must be zero, the coefficient of the x^2 term, so that if r is the third solution,

$$(a + b\sqrt{c}) + (a - b\sqrt{c}) + r = 0,$$

$$2a + r = 0,$$

$$r = -2a.$$

This contradicts the assumption that $a + b\sqrt{c}$ represented a number in the least-inclusive extension field of solutions. The conclusion is that the solutions of $8x^3 - 6x - 1 = 0$ are not constructible numbers, so the following theorem is established:

THEOREM 5.5. The construction of trisecting the general angle cannot be performed by use of the straightedge and compass alone.

The third and most complex of the three famous Greek problems is that of squaring a circle. This means to find the length of one edge of a square that has the same measure of area as that of a circle whose radius is known. Algebraically, if x is the required length of edge and if the length of the known radius is one, then $x^2 = \pi$. The length $\sqrt{\pi}$ is the required measure of the edge. Proving that this number cannot be constructed depends on the following two statements, both of which are true but not proved here.

1.   All constructible numbers are algebraic.

DEFINITION. Algebraic numbers are solutions of an algebraic equation of the form

$$a_n x^n + a_{n-1} x^{n-1} + \cdots + a_0 = 0$$

with integral coefficients, and with $n \geq 1$ and $a_n \neq 0$.

2.   T. Lindemann proved in 1882 that the number π is not an algebraic number, so $\sqrt{\pi}$ is not algebraic either.

THEOREM 5.6. The construction of squaring a circle is impossible by means of a straightedge and compass alone.

As mentioned earlier, the fact that the Greeks could not perform these three constructions by the use of the straightedge and compass did not keep them from finding solutions by other means. Many advances in mathematics, including much of the theory of conics, probably arose as the result of attempts to provide solutions for construction problems. Two examples showing how the constructions can actually be performed are taken from the history of mathematics. One of these is the use of a device or curve called the *conchoid of Nicomedes,* who lived about 240 B.C. The actual method of trisecting an angle by using the conchoid is illustrated in Figure 5.16a. The curve through G is the conchoid. For a

(a)   (b)

FIGURE 5.16

fixed line $\overleftrightarrow{CD}$ and a fixed point O not on $\overleftrightarrow{CD}$, the conchoid (which really consists of two branches, one on each side of $\overleftrightarrow{CD}$) is the set of points defined as follows: Consider the set of all lines through O and intersecting $\overleftrightarrow{CD}$. Take a fixed distance on the line beyond the point of intersection. For example, let $\overrightarrow{OB}$ be one of the lines, with EX the fixed distance. For a given point, line, and distance, the set of all points X the fixed distance from $\overleftrightarrow{CD}$ along the rays from O is a branch of a conchoid.

If the curve (or rather the mechanical instrument) is placed as in Figure 5.16, with $\overleftrightarrow{CD}$ perpendicular to $\overleftrightarrow{OA}$ and with $XE = 2(EO)$, then the given angle AOB can be trisected simply by locating point E on the line on $\overrightarrow{OB}$, drawing $\overline{EG}$ parallel to $\overleftrightarrow{OA}$, and connecting O to G, the point on the conchoid. $\overrightarrow{OG}$ is the trisector of the given angle.

The reason why $\overrightarrow{OG}$ trisects the angle depends on the theory explained in connection with Figure 5.16b. $GI = 2EO$ by assumption. If H is the midpoint of $\overline{GI}$, then $\overline{EO}$, $\overline{HI}$, $\overline{GH}$, $\overline{EH}$ are all congruent. Because of the isosceles triangles, $m\angle EOH = m\angle EHO = m\angle HGE + m\angle HEG = 2m\angle HGE$. But $\angle FOI \cong \angle HGE$ because of parallel lines, so $2m\angle HGE = 2m\angle FOI$. The result is that $\angle EOG$ has a measure twice that of $\angle FOI$, so $\overrightarrow{OI}$ is the trisector.

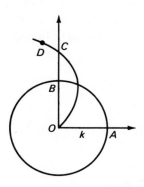

FIGURE 5.17

A second example of a curve from higher geometry used to construct a solution to one of the famous problems is the use of the *spiral of Archimedes* to solve the problem of squaring the circle. The spiral of Archimedes is the curve OCD shown in Figure 5.17, with polar equation $r = k\theta$, for k a given constant. If a circle has a radius k

as shown, then the arc for any angle AOB has length $k\theta$, which is also the length of $r = OC$. The area of the circle is

$$\pi k^2 = \frac{k}{2}(4\overset{\frown}{AB}).$$

If OA and OB are perpendicular, then

$$\frac{k}{2}(4\overset{\frown}{AB}) = 2k(OC).$$

If a square of side x is to have the same size as the circle, then

$$x^2 = 2k(OC),$$

and x can be constructed.

In addition to the three famous problems, the equally interesting problem of constructing a regular polygon inscribed in a given circle has concerned mathematicians since the time of the Greeks. In this case, the proof for the general problem was again algebraic and was provided by the great mathematician Karl Gauss. At the age of 18, he solved the previously unsolved problem of how to inscribe a regular polygon of 17 sides in a circle, using only a straightedge and compass. He also proved the theorem that tells which regular polygons can and cannot be inscribed. This theorem is stated here without proof.

THEOREM 5.7. A regular polygon can be inscribed in a circle by means of a straightedge and compass alone if and only if the number of sides, n, can be expressed as $2^x \cdot p_1 \cdot p_2 \ldots p_k$, for x a non-negative integer and each p_i a distinct prime of the form $2^{2^y} + 1$, for $y \geq 0$.

Some of the regular polygons that are constructible, according to the theorem of Gauss, are those with 3, 4, 5, 6, 8, 10, 12, 15, 16, 17, 20, or 24 sides. The statement of the theorem gives no clue as to how to prove the possibility or impossibility in any specific case without reference to the general theorem. The specific cases for $n = 10$ and $n = 7$ are discussed briefly.

Figure 5.18 shows the analysis for the regular decagon. The central angle AOB has a measurement of $36°$, and angles OAB and

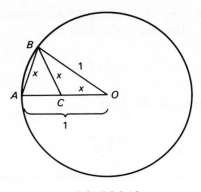

FIGURE 5.18

ABO each have measurements of $72°$. If $\overline{BC}$ is the bisector of $\angle ABO$, then triangles ABC and BCO are both isosceles, so $\overline{AB} \cong \overline{BC} \cong \overline{CO}$. Also,

$$\triangle ABO \sim \triangle ACB,$$

and

$$\frac{1}{x} = \frac{x}{1-x},$$

or $x^2 + x - 1 = 0$. The positive solution is

$$x = \frac{\sqrt{5} - 1}{2},$$

which is a constructible number. You should recognize x as the golden ratio.

As Theorem 5.7 indicates, the regular decagon can be constructed in a circle, whereas the regular heptagon (seven sides) cannot. The proof that it is impossible to construct a heptagon in a circle is similar to the proof concerning the trisection of a general angle. The problem is equivalent to constructing a length $x = 2\cos 2\pi/7$. If $2\pi/7 = \theta$, then

$$3\theta + 4\theta = 360° \quad \text{and} \quad \cos 3\theta = \cos 4\theta.$$

Trigonometric identities can be used to express these cosines in terms of x.

$$2\cos 3\theta = 2(4\cos^3\theta - 3\cos\theta) = x^3 - 3x$$
$$2\cos 4\theta = 2(2\cos^2 2\theta - 1) = 4(2\cos^2\theta - 1)^2 - 2$$
$$= (x^2 - 2)^2 - 2$$

Setting the two expressions equal yields a quartic equation in x.

$$x^4 - x^3 - 4x^2 + 3x + 2 = 0,$$

which can be factored as $(x - 2)(x^3 - 2x - 1) = 0$. It is left as an exercise to show why these factors do not give values for x that can all be constructed with straightedge and compass.

EXERCISES 5.4

1. Apply the definition to show that $\sqrt{2 + \sqrt{2}}$ is an algebraic number.

2. Is $\sqrt[3]{2}$ an algebraic number? Why?

3. Find an approximate decimal solution for the length of an edge of a cube that has twice the volume of another cube with an edge 2 inches long.

4. Prove that if a number of the form $a + b\sqrt{c}$ is a solution of $x^3 + dx^2 + ex + f = 0$, so is $a - b\sqrt{c}$.

5. Explain why the equation $8x^3 - 6x - 1 = 0$ has no rational solutions.

6. Explain how a mechanical device might be made that would actually draw one branch of a conchoid.

7. Show that the spiral of Archimedes can also be used for trisecting an angle.

8. Use Theorem 5.7 to list those regular polygons with between 24 and 30 sides that can be constructed in a circle.

9. In the quartic equation for the regular heptagon, explain why the first factor cannot be used to find a solution.

10. In the quartic equation for the regular heptagon, explain why the second factor cannot be used to find a solution.

5.5 CONSTRUCTIONS BY PAPER FOLDING

As discussed earlier in this chapter, the Greeks specified that geometric constructions should be performed by straightedge and compass alone; nevertheless, they devised other instruments when some problems proved impossible with the given instruments. Since that time, other ways of accomplishing the same constructions have been devised. One that may seem elementary, yet has occupied the attention of mathematicians in recent years, is the approach through *paper folding.* In actual practice, waxed paper is ordinarily used so that the crease will remain visible. A few of the basic constructions are illustrated in this section so that you will catch the spirit of this work and thus be able to contrast it with other construction methods in Euclidean geometry.

   Folding a straight line that is the perpendicular bisector of a given segment. Suppose $\overline{AB}$ is given, as in Figure 5.19a. Fold the paper so that point A is superimposed on point B and crease, as in Figure 5.19b. $\overleftrightarrow{CD}$ is the required line.

(a)   (b)

FIGURE 5.19

   Folding the bisector of a given angle (Figure 5.20). Fold the paper with the crease through the vertex and with one side of the angle superimposed on the other.

FIGURE 5.20

   Folding a line through a given point and parallel to a given line. First fold a crease for $\overleftrightarrow{CD}$ perpendicular to the given line $\overleftrightarrow{AB}$ (Figure 5.21), then fold a crease through the given point G for a line $\overleftrightarrow{EF}$ perpendicular to $\overleftrightarrow{CD}$. That line is parallel to $\overleftrightarrow{AB}$ and is the required line.

FIGURE 5.21

Folding a perpendicular from a point to a line. Simply fold a perpendicular to the line through the point by superimposing part of line $\overleftrightarrow{AB}$ on itself (Figure 5.22).

FIGURE 5.22

Although paper folding seems very simple, mathematicians have been able to prove the following rather startling theorem about paper-folding constructions:

THEOREM 5.8. All of the constructions of the plane Euclidean geometry that can be performed by straightedge and compass can also be performed by folding and creasing paper.

This remarkable theorem is based on several assumptions that can be found in the excellent reference on paper folding, *Paper Folding for the Mathematics Class,* by Donovan A. Johnson. These include, for example, the assumption that paper can be folded in such a way that one line can be superimposed on another line on the same sheet of paper and that the crease formed is in fact a straight line. Even the brief analysis of paper folding in this section gives additional insight into the nature of the approximations involved in ordinary constructions. The theory of paper folding is just as mathematical and just as exact as the theory of constructions with straightedge and compass, but the methods differ widely.

Paper folding can be used for a somewhat different type of construction not usually studied in ordinary college geometry, that of constructing tangents to a parabola. Figure 5.23a illustrates how a series of tangents to the parabola can be determined by folding the focus F

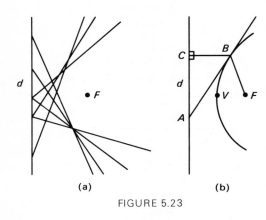

(a)   (b)

FIGURE 5.23

onto the directrix d. The reason why this construction works is explained in Figure 5.23b. It can be proved that the tangent $\overleftrightarrow{BA}$ at a point on the parabola is the angle bisector of $\angle FBC$, where F is the focus and C is the foot of the perpendicular from B to the directrix.

EXERCISES 5.5

Using waxed paper, perform the following constructions for Exercises 1–9.

1. Fold a straight line that is the perpendicular bisector of a given segment.
2. Construct the bisector of a given angle.
3. Construct a line through a given point and parallel to a given line not through the point.
4. Construct the perpendicular from a point to a line not through the point.
5. Construct ten tangents to a parabola, given the directrix and the focus.
6. Construct the tangent at a point on a circle.
7. Construct the incenter of a given triangle.
8. Construct the circumcenter of a given triangle.
9. Construct the orthocenter of a given triangle.

10. Prove that the paper-folding method of constructing lines tangent to a parabola is valid.

5.6   CONSTRUCTIONS WITH ONLY ONE INSTRUMENT

In every section of this chapter, including Section 5.5 on paper folding, the concepts introduced have been closely related to the Greek restrictions of using only the compass and the straightedge to perform constructions. A somewhat radical departure is to determine whether it is really necessary to use both instruments or whether it might be possible, using just one instrument, to perform all of the constructions that can be performed with both. For students who have trouble making constructions with both instruments available, this may seem sheer folly, yet mathematicians are interested in finding the truly minimum set of instruments necessary for making constructions.

Historically, attempts to discover the constructions possible with the collapsing compass alone came first. However, constructions using the straightedge alone will be discussed first here. Though the mathematical interest in this problem comes from the middle ages, it was not until the invention of projective geometry in the early nineteenth century that the theory was established on a firm foundation. The basic result is stated in what is known as the Poncelet-Steiner construction theorem.

THEOREM 5.9. *Poncelet-Steiner construction theorem.* All of the constructions that can be performed with the straightedge and compass can be performed with the straightedge alone, given a single circle and its center.

As Theorem 5.9 indicates, it is not possible to perform all of the constructions of Euclidean geometry using just the straightedge. But it is possible if one circle and its center are also given. This implies that the length of the radius and the midpoint between two ends of the diameter are known. Despite the use of the one circle, many of the concepts in this theory are from projective geometry (a geometry that does not include the circle and its radius as invariants, as will be seen in Chapter 7).

As explained in the previous paragraphs, the attempt to limit the instruments used in constructions to the straightedge alone was not totally successful, since a single circle and its center were needed. The second alternative is to limit the instrument to the compass alone. Obviously, it is impossible to draw a straight line with the compass alone, so it must be understood that a line is completely determined if two points on it are found or given. Constructions with the compass alone are called *Mohr-Mascheroni constructions*. C. Mohr published the first known account of these constructions in 1672, although his book was not well known to mathematicians until 1928 when it was rediscovered. Meanwhile, the Italian mathematician Mascheroni, during the last half of the eighteenth century, had independently discovered the following theorem:

THEOREM 5.10. All constructions possible by use of the straightedge and compass can also be made by use of the compass alone.

For a discussion of this theorem, see H. Rademacher and O. Toeplitz, *The Enjoyment of Mathematics.*

The following example illustrates a construction performed entirely with the compass.

Find the midpoint of a given segment by use of a compass alone.

1.   Consider $\overline{AB}$, as in Figure 5.24. Construct the circle with B as center and BA as radius.

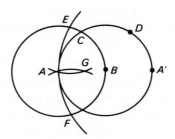

FIGURE 5.24

2.   With AB as radius, mark off three arcs $\overset{\frown}{AC}$, $\overset{\frown}{CD}$, $\overset{\frown}{DA'}$, locating A' so that AA' is a diameter of the circle.
3.   Construct the circle with center A and radius AB.

4. Construct the circle with center A' and radius $A'A$, intersecting the circle of step 3 at the points E and F.
5. Construct the circles with centers E and F and radius EA. These meet at A and the required point G.

The proof that this rather elaborate construction results in the correct point is based on the fact that A' and G are *inverse points* with respect to the circle with center A and radius AB. The geometry of inverse points is discussed in the following chapter, but the proof used here does not require a knowledge of that concept.

Triangles $A'EA$ and EGA are similar, so

$$\frac{AA'}{AE} = \frac{AE}{AG}, \quad \text{or} \quad AA' \cdot AG = (AE)^2.$$

Because

$$AA' = 2AB = 2AE,$$

$$2AE \cdot AG = (AE)^2,$$

$$2AG = AE = AB,$$

and G is the midpoint of $\overline{AB}$, as was to be proved.

The proof of Theorem 5.10 consists of showing how to find the points of intersection of a straight line and a circle and the points of intersection of two straight lines with the compass alone, since it is obviously simple enough to draw a circle and to find the points of intersection of two circles with the compass alone. The method of *constructing the points of intersection of a circle and a line not through the center* is shown in Figure 5.25. Let B and C be the given points, with A the center of the given circle.

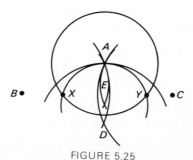

FIGURE 5.25

1. Construct circles with centers B and C, the given points on the line, passing through A and intersecting again in D.

2. Following steps 4 and 5 of the previous construction, find a point E such that $AD \cdot AE = r^2$ for r the radius of the original circle. (This construction must be modified if D is inside the original circle.)

3. The circle with center E and radius EA intersects the original circle in the required points X and Y.

The proof that X and Y are the points of intersection of the given line and circle is left as Exercise 1, Exercise Set 5.6.

This chapter has shown that constructions in Euclidean geometry are important in that they provide additional information about concepts previously studied. In the last section, the somewhat surprising conclusion is that constructions also provide one transition from Euclidean geometry to the geometry of inversion and to projective geometry, which will be studied in the next two chapters. This study will involve other relationships between these geometries and Euclidean geometry, including a classification based on transformations.

EXERCISES   5.6

1. Complete the proof of the Mascheroni construction of the point of intersection of a circle and a line not through the center.

With the compass alone, perform the following Mohr-Mascheroni constructions and briefly describe each step. It is assumed that a line is given by two distinct points on the line.

2. Given a circle and a point inside the circle but distinct from the center, find a point collinear with the center and the given point such that the product of the distances of the two points from the center is equal to the square of the radius of the given circle.

3. Find a distinct point C on $\overleftrightarrow{AB}$ such that $AB = BC$, if A and B are given points.

4. Find the midpoint of a given arc of a circle.

5. Construct the perpendicular to a given line at a given point on the line.

6. Construct the parallel to a given line through a point in the plane but not on the line.

For Exercises 7–11, give the proof of the constructions performed previously in the indicated exercises.

7. Exercise 2.

8. Exercise 3.

9. Exercise 4.

10. Exercise 5.

11. Exercise 6.

CHAPTER 6

THE GEOMETRY OF INVERSION

6.1 BASIC CONCEPTS

With the exception of the finite geometries of Chapter 1, the previous chapters have been an expansion of the relatively familiar geometry of Euclid. In the last four chapters of the text, new geometries with some properties unlike those of Euclidean geometry are discussed. The first of these is the geometry of inversion.

The geometry of inversion begins with the definition of inverse points.

DEFINITION. Two points P and Q are *inverse points* with respect to a fixed collinear point O on the line if $OP \cdot OQ = c$, for c a positive real number.

Both distances are measured in the same direction from O. The circle with center O and radius $\sqrt{c}$, shown in Figure 6.1, is the

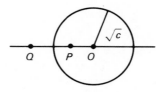

FIGURE 6.1

circle of inversion. The point O is called the *center of inversion.* From these statements, one may conclude that the product of the distances of two inverse points from the center of inversion is equal to the square of the radius of the circle of inversion, called the *radius of inversion.* In this chapter, distances are always considered positive.

In its simplest interpretation, the concept of inversion is numerical in nature, consisting merely of two distances that always have the same product. From this intuitive statement, it is easy to see that if one of the points moves away from the center, the second must move toward it. A numerical example will illustrate this arithmetic interpretation. Suppose that point P (in Figure 6.1) is at a distance of two units from the point O and that the constant product, called the *constant of inversion,* is 9. Then

$$2 \cdot OQ = 9,$$

and

$$OQ = 9/2.$$

Now suppose that the constant is unchanged but that $OP = 1$. Then $OQ = 9$. If the radius of inversion is 1, then the numerical values of the distances are reciprocals, and one can be found from the other by inverting the number. The name "inversion" may be related to this fact.

The concept of inversion makes it possible to pair the points on a line in reference to one fixed point on the line with a constant that represents the square of the radius of the circle of inversion. In this pairing of points, two special cases become apparent. A point on the circle of inversion is its own inverse point. In other words, these points are *self-inverse*. The second special case is that no point on the line is the inverse of the center of inversion, since division by zero does not result in a real number. This last statement shows the need for a consideration of inversion from a more general point of view.

A review of the definition of transformation given in Chapter 2 shows that inversion as defined so far is not a transformation, since there is an exception to the pairing. But if the *inversive plane* is created by including on each line an inverse point for the center of inversion, then the definition of transformation is satisfied. A line in the inversive plane is formed by including, in addition to the ordinary points, a special point called an *ideal point*, defined to be the image of the center of inversion. With the addition of the ideal point, each point on the line of inversive geometry has a unique image under inversion. The line of inversive geometry is not an ordinary Euclidean line. The ideal point in inversive geometry is not a concept from Euclidean geometry but a new concept created to guarantee a desired one-to-one correspondence of points.

Because of the fact that the center of inversion in the inversive plane is to have a unique image in the inversive plane, then each line through the center must have the same inverse point for it, and the new point created must lie on each line through the original point. The inversive plane is illustrated symbolically in Figure 6.2, with O the center

FIGURE 6.2

of inversion and with I (the ideal point on the inversive plane) the inverse of the center on any line through O. It should be understood thoroughly that the inversive plane, because it includes the one ideal point that lies on each line through the center of inversion, has some properties unlike the Euclidean plane with which you are familiar. The transformation of inversion in the inversive plane is a new transformation that can be studied in its own right. As usual, it is good to speculate about these matters before reading on.

The invention of the transformation of inversion is sometimes credited to L. J. Magnus in 1831. But prior to this time, Vieta in the sixteenth century and Robert Simson in the eighteenth century were aware of elements of the theory. Some mathematicians give the major credit to Steiner, but his work on the subject was not published. Many mathematicians worked during the 1830s and 1840s to develop the general theory further.

The basic properties of the transformation of inversion depend on the peculiar properties of the inversive plane, a plane with one ideal point. It should be understood that the inversive plane as used here includes real points only and not the complex points that will be introduced later.

THEOREM 6.1. The circle of inversion is an invariant under the transformation of inversion.

This theorem follows from the fact that each point of the circle of inversion is its own inverse under inversion with respect to that circle. The circle is pointwise invariant, according to the following definition.

DEFINITION. A set of points is *pointwise invariant* for a transformation if every point is its own image.

THEOREM 6.2. The inverse of a line through the center of inversion is the same line.

This line is not pointwise invariant because, in general, a second distinct point on the line is the image of a given point on the line. The fact that the inverse points and the center of inversion are collinear is actually a part of the definition of inverse points.

A more significant theorem, and possibly one that may not have been expected, concerns the inverse curve for a straight line not passing through the center of inversion in the inversive plane.

THEOREM 6.3. The image of a straight line not through the center of inversion, under the transformation of inversion, is a circle passing through the center of inversion.

In Figure 6.3, let P and P' and Q and Q' be pairs of inverse points with respect to the circle of inversion with center O. The given line whose image is desired is $\overleftrightarrow{PQ}$, and $\overrightarrow{PO}$ is perpendicular to $\overline{PQ}$.

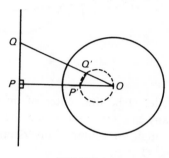

FIGURE 6.3

By the definition of inverse points,

$$OP \cdot OP' = OQ \cdot OQ',$$

or

$$\frac{OP}{OQ} = \frac{OQ'}{OP'}.$$

This proportion implies that $\triangle OPQ \sim \triangle OQ'P$ because they have an angle in common. Thus $\triangle OQ'P$ is a right triangle, with the vertex of the right angle at Q'. If Q is considered a variable point on the given line, then Q' is a variable vertex of a right triangle inscribed in a semicircle, with $\overline{OP'}$ the diameter of the circle. The proof of the converse of Theorem 6.3 consists of reversing the steps in the previous argument and is left as an exercise.

The fact that the image of a line under inversion is sometimes a circle rather than a line shows that the property of being a straight line is not always an invariant property under the transformation of inversion. This is a major difference between the geometry of inversion and ordinary Euclidean geometry. The relationship between circles and lines under inversion can be further extended by investigating the inverse of a circle not passing through the center of inversion. Before reading ahead, you should predict what this image will be.

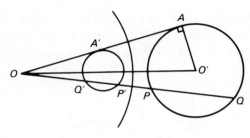

FIGURE 6.4

In Figure 6.4, let O be the center of inversion, with circle O' the given circle. For any line $\overleftrightarrow{OP}$ intersecting the circle in two distinct points and passing through the center of inversion, if P' and Q' are the inverse points for P and Q, then

$$OP \cdot OP' = OQ \cdot OQ' = c,$$

the constant of inversion. Also

$$OP \cdot OQ = (OA)^2 = k,$$

a constant equal to the square of the length of the tangent from O to the given circle.

$$\frac{OP'}{OQ} = \frac{OQ'}{OP} = \frac{c}{k},$$

which is a constant.

The situation reduces to a familiar transformation, a homothetic transformation with center O. This means that, when P is a point on the given circle, the image P' is a point on a circle that is homothetic to the given circle; thus c/k is the homothetic ratio. This completes the proof of the next theorem.

THEOREM 6.4. The image under inversion of a circle not passing through the center of inversion is a circle.

FIGURE 6.5

This section concludes with one other important property of the transformation of inversion. Suppose, as in Figure 6.5, that P and P' are inverse points with respect to a given circle with center O. The transformation takes P to its inverse point P'. If the same inversive transformation is applied again, the image of P' is P. The result is that the product of an inversion followed by the same inversion is the identity. The inversion transformation is called an *involutory transformation*, or a *transformation of period two*, because two successive applications result in the identity. It should be observed that an inversion transformation is its own inverse, which provides another good reason for the name given to the correspondence.

EXERCISES   6.1

Complete the following table:

	Distance from center of inversion to original point	Radius of circle of inversion	Distance of inverse point from center of inversion
1.	3	4	?
2.	3	3	?
3.	3	2	?
4.	3	1	?
5.	9/8	3/7	?
6.	2/3	?	5/6
7.	?	9/5	6/7

8. Do each two points in the inversive plane determine a unique line?

9. Describe the location of the inverse of any point inside the circle of inversion.

10. When is the image of a circle under inversion a circle, and when is it not?

11. Prove the converse of Theorem 6.3.

12. What is the product of inversive transformations if the same transformation is used as a factor an even number of times?

13. What are possible images under inversion for a set of points consisting of two distinct intersecting lines?

14. Describe the inverse of a circle outside the circle of inversion and concentric to it.

15. Suppose a curve intersects its inverse curve. Where are the points of intersection?

16. In Section 5.6, the construction was given to find the midpoint of a given segment by use of a compass alone. Use the concept of inversion to prove that this construction results in the required point.

6.2   ADDITIONAL PROPERTIES AND INVARIANTS UNDER INVERSION

In the previous section, it was found that the circle of inversion is pointwise invariant under inversion and that a line through the center of inversion is invariant, but not pointwise invariant. A set of points is pointwise invariant if each point is its own image under a transformation. It was further found that coincidence of points and curves is preserved under inversion. These facts provide the basis for the continued investigation of invariant properties under the inversion transformation. One of the most useful of the invariants under inversion is indicated in the next theorem.

THEOREM 6.5. The measure of the angle between two intersecting curves is an invariant under the transformation of inversion.

The meaning of this theorem is shown symbolically in Figure 6.6. The angle between two curves C_1 and C_2 is defined to be the angle between their tangents at the point of intersection. One of the two supplementary angles must be designated. The images of the two inter-

FIGURE 6.6

secting curves are two intersecting curves C_1' and C_2'. The angle between their two tangents at the point of intersection has the same measure as between the two original tangents.

The proof of Theorem 6.5 depends on proving that the angle that a curve makes with a line through the center of inversion is congruent to the angle the inverse curve makes with the same line.

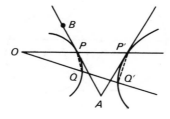

FIGURE 6.7

In Figure 6.7, let curve PQ and $P'Q'$ be inverse curves with O the center of inversion and with P, P' and Q, Q' pairs of inverse points. The four points P, P', Q', Q all lie on a circle (the proof of this statement is left as an exercise), so opposite angles are supplementary. Thus,

$$\angle OPQ \cong \angle P'Q'O.$$

Applying a fundamental concept from analysis, one may conclude that the limiting position of $\overrightarrow{OQ}$ is $\overrightarrow{OP}$, as Q approaches P along the original curve. The tangents $\overleftrightarrow{PA}$ and $\overleftrightarrow{P'A}$ are the limiting positions of the secants $\overleftrightarrow{PQ}$ and $\overleftrightarrow{P'Q'}$. Thus, the limiting position of the angle from $\overrightarrow{OP}$ to the original curve is $\angle OPB$, which is congruent to $\angle OP'A$, the limiting position of the angle from $\overrightarrow{OP}$ to the image curve.

It is important to observe that, although the measure of the angle is preserved under inversion, the direction of the angle is reversed. In Figure 6.7, the angle from $\overrightarrow{OP}$ to $\overrightarrow{PB}$ is measured in a clockwise direction, whereas the angle from $\overrightarrow{OP'}$ to $\overrightarrow{P'A}$ is measured in a counterclockwise direction. For this reason, inversion (like reflection) is another example of an opposite transformation.

As a corollary to Theorem 6.5, two curves that intersect at right angles have images that also meet at right angles. *Orthogonal circles,* two circles intersecting at right angles, play a very important role in the theory of inversion. A fundamental concept is included in the next theorem.

THEOREM 6.6. A circle orthogonal to the circle of inversion is an invariant set of points (but not pointwise invariant) under the inversion transformation.

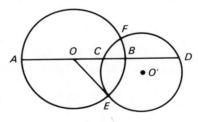

FIGURE 6.8

In Figure 6.8, circle O is the circle of inversion and circle O' is any circle orthogonal to it. If a line through O intersects circle O' at C and D, then $OC \cdot OD = (OE)^2$, for E a point of intersection of the two circles. But since $\overline{OE}$ is also the radius of the circle of inversion, C and D are by definition inverse points with respect to the circle of inversion.

For each point of $\overset{\frown}{FDE}$, the image under inversion is a point of $\overset{\frown}{FCE}$, with the endpoints F and E invariant points. Thus, circle O' is a set of invariant points.

A second useful theorem about orthogonal circles under inversion is the following:

THEOREM 6.7. If two circles orthogonal to the circle of inversion intersect each other in two points, these two points are inverse points.

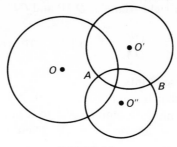

FIGURE 6.9

In Figure 6.9, let O be the circle of inversion with O' and O'' two circles orthogonal to it and intersecting each other at A and B. It can be proved that point O is collinear with A and B because it is a point from which congruent tangents can be drawn to circle O' and O''. (The proof of this statement is left as an exercise.) Then, from Theorem 6.6, A and B are inverse points.

Recall that the concept of inverse point was encountered briefly in the last section of Chapter 5 on construction. Speculation about how a point inverse to a given point might actually be constructed could lead to discovery of an additional fundamental property of inverse points that is implied in the next theorem.

THEOREM 6.8. The inverse point for a point outside a circle lies on the line joining the points of intersection of the tangents from the point to the circle.

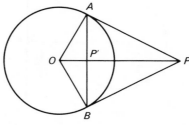

FIGURE 6.10

In Figure 6.10, P is the external point. We can prove that P and P' are inverse points because it is known by definition that the inverse of P lies on $\overleftrightarrow{OP}$.

From similar right triangles OAP and $OP'A$,

$$\frac{OA}{OP'} = \frac{OP}{OA}, \qquad (OA)^2 = OP \cdot OP',$$

so P and P' are inverse points.

From Theorem 6.8 and Figure 6.10, it is possible to devise a method of constructing the point inverse to a given point, whether the given point is inside or outside the circle. These construction problems are included in the exercises.

DEFINITION. In inversive geometry, the line through an inverse point and perpendicular to the line joining the original point to the center of the circle of inversion is called the *polar* of the original point, whereas the point itself is called the *pole* of the line.

In Figure 6.10, $\overleftrightarrow{AB}$ is the polar of P and P is the pole of $\overleftrightarrow{AB}$. Recall that the words pole and polar were encountered in a different setting in one of the finite geometries of Chapter 1.

In Figure 6.11, if P and P' are inverse points, then $\overleftrightarrow{P'Q'}$ is the polar of P and P is the pole of $\overleftrightarrow{P'Q'}$. From Figure 6.11, you might surmise that the reciprocal relation between poles and polars exists as stated formally in the next theorem.

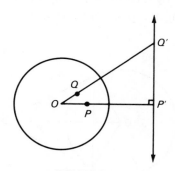

FIGURE 6.11

THEOREM 6.9. If a second point is on the polar of a first, with respect to a given circle of inversion, then the first is on the polar of the second with respect to the same circle.

In Figure 6.11, if $\overleftrightarrow{P'Q'}$ is the polar of P, then it must be shown that the polar of Q' passes through P. If Q is the inverse point for Q', then all that is needed is to show that $\overleftrightarrow{QP}$ is perpendicular to $\overrightarrow{OQ'}$. The two pairs of inverse points, P, P' and Q, Q', lie on a circle. Because $\angle PP'Q'$ is a right angle, $\angle Q'QP$, the opposite angle in the inscribed quadrilateral, is also a right angle. The theory of poles and polars is also important in projective geometry. Some extensions of the theory are found in the following exercises.

EXERCISES   6.2

1. List the possibilities for the images under inversion of two lines meeting at right angles at a point other than the center of inversion.

2. In the proof of Theorem 6.5, prove that P, P', Q', Q all lie on a circle.

3. In the proof of Theorem 6.7, prove that point O is collinear with points A and B.

4. Given a point outside a given circle, construct its inverse point with respect to the circle.

5. Given a point inside a given circle, construct its inverse point with respect to the circle.

6. Prove that the polars of every point on a line are concurrent at the pole of the line.

7. Construct the pole of a line not intersecting the circle of inversion.

8. Construct the pole of a line intersecting the circle of inversion in two distinct points.

9. A triangle is *self-polar* with respect to the circle of inversion if each side is the polar of the opposite vertex. Explain how to construct a triangle self-polar to a given circle.

10. Describe the inverse of a circle with respect to a point on the circle.

11. Through two points inside a given circle but not collinear with the center, construct a circle orthogonal to the given circle.

6.3   THE ANALYTIC GEOMETRY OF INVERSION

For the Euclidean motions, it was possible to write a set of equations that clearly showed the one-to-one correspondence. In that

case, the transformations were expressed in terms of a set of simultaneous linear equations. The transformation of similarity was also
described algebraically using sets of linear equations. The transformation
of inversion can also be approached from an analytic point of view, and
this approach will give additional insight into the nature of the
transformation.

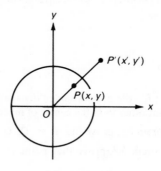

FIGURE 6.12

The equations for inversion can be developed using Figure 6.12.
The circle with center at the origin and with radius r can be considered the center of inversion. Inverse points P and P' have coordinates
(x, y) and (x', y'), respectively.

Since $OP \cdot OP' = r^2$,

$$\left(\sqrt{x^2 + y^2}\right)\left(\sqrt{x'^2 + y'^2}\right) = r^2,$$

or

$$(x^2 + y^2)(x'^2 + y'^2) = r^4.$$

Because P and P' are collinear with O, $x'/x = y'/y$, $x'y = xy'$,
or $x'^2y^2 = x^2y'^2$; therefore,

$$(x^2 + y^2)\left(\frac{x^2y'^2}{y^2} + y'^2\right) = r^4,$$

$$\frac{1}{y^2}(x^2 + y^2)(x^2y'^2 + y'^2y^2) = r^4,$$

$$\frac{y'^2}{y^2}(x^2 + y^2)(x^2 + y^2) = r^4.$$

But

$$\frac{y'^2}{y^2} = \frac{x'^2}{x^2},$$

and if both sides are multiplied by x^2, then $x'^2(x^2 + y^2)^2 = x^2r^4$. Taking the square root of both sides shows that

$$x'(x^2 + y^2) = xr^2$$

or

$$x' = \frac{xr^2}{x^2 + y^2}.$$

Similarly, the expression for y' can be found so that the next theorem is proved.

THEOREM 6.10. The equations for the inversion transformation of a point $P(x, y)$ into $P'(x', y')$ relative to the circle $x^2 + y^2 = r^2$ are

$$x' = \frac{xr^2}{x^2 + y^2}, \qquad y' = \frac{yr^2}{x^2 + y^2}.$$

The equations in Theorem 6.10 show clearly that the image of the center of inversion is the ideal point in the inversive plane, since $x^2 + y^2 = 0$, and x' and y' do not have real values. They also show that for points on the circle of inversion, $x = x'$ and $y = y'$.

EXAMPLE. Find the inverse of $(2, 3)$ with respect to the circle $x^2 + y^2 = 9$.

$$x' = \frac{9(2)}{4 + 9} = \frac{18}{13}, \qquad y' = \frac{9(3)}{13} = \frac{27}{13}$$

The inverse point is

$$\left(\frac{18}{13}, \frac{27}{13}\right).$$

Solving the equations in Theorem 6.10 for x and y provides an analytic proof of the fact that the inverse transformation is its own inverse. The result is that x and x' can be interchanged and y and y' can be interchanged so that

$$x = \frac{x'r^2}{x'^2 + y'^2}, \qquad y = \frac{y'r^2}{x'^2 + y'^2}.$$

The requirement that the circle of inversion have its center at the origin is of course not mandatory. The substitution

$$x'' = x + h, \qquad y'' = y + k$$

shows that more general forms of the equations for inversion are

$$x' = \frac{(x'' - h)r^2}{(x'' - h)^2 + (y'' - k)^2}, \qquad y' = \frac{(y'' - k)r^2}{(x'' - h)^2 + (y'' - k)^2}.$$

It is instructive to use the analytic formulas to verify the inversion theorems of the last two sections for particular numerical examples. For example, what is the image of the line $x - 4 = 0$ under inversion with respect to a circle with center at the origin and radius three? (See Figure 6.13.)

The equation of the image is obtained by substituting

$$\frac{9x'}{x'^2 + y'^2}$$

for x in the equation of the line. Thus,

$$\frac{9x'}{x'^2 + y'^2} - 4 = 0,$$

$$\frac{9x'}{x'^2 + y'^2} = 4,$$

$$9x' = 4(x'^2 + y'^2),$$

or

$$4x'^2 - 9x' + 4y'^2 = 0.$$

This is a circle through the origin, as expected.

The equations of the transformation of inversion and its inverse transformation also make it possible to find the equations for the inverse of curves not covered in previous theorems. An example

FIGURE 6.13

illustrates the wide variety of possibilities. Find the image of the parabola $y^2 = 4x$ under inversion with respect to the circle $x^2 + y^2 = 4$. (See Figure 6.14.)

FIGURE 6.14

The substitution for x and y in the equation of the parabola results in

$$\frac{(4y')^2}{(x'^2 + y'^2)^2} = 4\frac{4x'}{x'^2 + y'^2}.$$

This equation can be somewhat simplified, as follows:

$$16y'^2 = 16x'(x'^2 + y'^2),$$

$$y'^2 = x'^3 + x'y'^2,$$

$$y'^2 - x'y'^2 = x'^3,$$

$$y'^2 = \frac{x'^3}{1 - x'}.$$

The inverse of the parabola in this case is not a parabola, nor even a conic, so the property of being a conic is not an invariant in inversive geometry.

When the study of inversion is extended from the real plane to the complex plane, a rather unexpected simplification appears in the equation of the transformation.

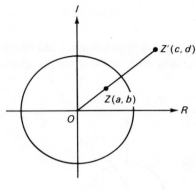

FIGURE 6.15

In Figure 6.15, let the real and imaginary axes be given. If the circle of inversion is a unit circle, and if $z' = c + di$ is the inverse of $z = a + bi$ with respect to this circle, then the following theorem shows the relationship between z and z'.

THEOREM 6.11. The transformation of points in the complex plane, with respect to the unit circle with center at the origin, has the equation $z' = 1/\bar{z}$ where $\bar{z}$ is the conjugate of z. If $z = a + bi$, then

$$z' = \frac{1}{\bar{z}} = \frac{1}{a - bi} \cdot \frac{a + bi}{a + bi}$$

$$= \frac{a + bi}{a^2 + b^2} \cdot$$

To show that z and z' are inverse points with respect to the unit circle, it is necessary to:

1. Show that the product of their distances from the origin is one.
2. Show that they are collinear with the origin.

Demonstrating that z and z' are inverse points is left as an exercise.
Inversion with respect to the circle with center at the origin and radius in the complex plane is accomplished by the transformation

$$z' = \frac{r^2}{\bar{z}}.$$

It is somewhat surprising that the numerical work of finding an inverse point in the complex plane is considerably simpler than is finding it in the real plane. For example, the inverse of $3 + 4i$, with respect to the circle with center at the origin and radius two, is $4/(3 - 4i)$, which can be simplified as follows:

$$\left(\frac{4}{3 - 4i}\right)\left(\frac{3 + 4i}{3 + 4i}\right) = \frac{4(3 + 4i)}{9 + 16} = \frac{12}{25} + \frac{16}{25}i.$$

EXERCISES   6.3

Use the equations for inversion with respect to a circle with center at the origin to find the inverse of the following real points.

	Point	Radius of inversion
1.	(3, 4)	1
2.	(1, 1)	2
3.	(0, 3)	3
4.	(7, 1)	4

For Exercises 5 and 6, find the inverse of the given point with respect to a circle with center (2, 3).

	Point	Radius of inversion
5.	(6, 2)	2
6.	(5, 5)	3

7. Find the equations for the inverse of the inversive transformation by actually solving for x and y in the original equations.

For each set of points given in Exercises 8–13, find the image under inversion with respect to the circle $x^2 + y^2 = 9$.

8. $x - 2 = 0$ 9. $x - 5 = 0$

10. $x + y + 2 = 0$ 11. $y^2 = 4x$

12. $x^2 = y$ 13. $x^2 + y^2 = 16$

14. Complete the proof of Theorem 6.11.

In the complex plane, find the inverse of the points given in Exercises 15–18, with reference to a circle with center at the origin and with the given radius.

	Point	Radius of inversion
15.	$2 - 5i$	1
16.	$3 + 7i$	1
17.	$1 - 2i$	2
18.	$3 + 11i$	3

6.4   SOME APPLICATIONS OF INVERSION

Three of the most apparent applications of inversion have already been mentioned but they need to be listed again before we proceed to new areas.

1. The theory of poles and polars is used in projective geometry.
2. Mascheroni constructions employ constructions of the inverse point as a basic technique in order to carry out many other constructions.
3. The finite geometry of Desargues (Chapter 1) used the concept of pole and polar.

A new fact about inverse points shows an application to an earlier construction problem.

THEOREM 6.12. Two inverse points with respect to a circle divide the diameter on which they lie internally and externally in the same ratio.

In Figure 6.16, $OC \cdot OD = (OB)^2$, or

$$\frac{OD}{OB} = \frac{OB}{OC}.$$

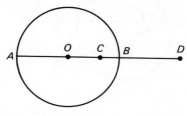

FIGURE 6.16

By a theorem on proportions, if $a/b = c/d$, then

$$\frac{a + b}{a - b} = \frac{c + d}{c - d},$$

so that

$$\frac{OD + OB}{OD - OB} = \frac{OB + OC}{OB - OC},$$

or

$$\frac{AD}{BD} = \frac{AC}{CB},$$

so that the ratios of division of $\overline{AB}$ by C and D are equal (directed segments have not been considered here). Note also that points A and B divide $\overline{CD}$ internally and externally in the same ratio; thus, the circle of inversion is the circle of Apollonius with respect to the two points C and D.

The fact that an inversion transformation sometimes transforms a straight line into a circle, or a circle into a straight line, is the basis for physical applications of inversion in linkages that change linear motion into curvilinear motion or vice-versa. The harnessing of the tides, the driving of a locomotive, and the production of electricity are all illustrations of one type of motion changed into another type of motion. The mechanical problem of devising linkages to do this sort of thing has concerned engineers during the past hundred years. One such device is

Peucellier's cell, whose construction is based on the theory of inversion. The device is illustrated in Figure 6.17.

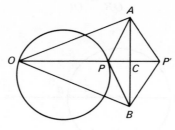

FIGURE 6.17

The quadrilateral $AP'BP$ is a rhombus, so $\overleftrightarrow{OC}$ is the perpendicular bisector of $\overline{AB}$.

$$OP \cdot OP' = (OC - PC)(OC + CP')$$
$$= (OC)^2 - (PC)^2$$
$$= (OA)^2 - (AC)^2 - [(PA)^2 - (AC)^2]$$
$$= (OA)^2 - (PA)^2$$

But in the mechanical device, which consists of the six rods forming the rhombus as well as $\overline{OA}$ and $\overline{OB}$, the distances OA and PA are constant, and thus P and P' are inverse points. If point P moves around the circle, P' traces a straight line.

Quite a different application of inversion is that of finding new theorems by inverting familiar ones. More precisely, the figure for a familiar Euclidean theorem, when inverted, may yield a new theorem by suggesting the corresponding properties. Even further, the new theorem does not have to be proved from the beginning; it is true because of the proof of the original theorem and the accepted properties of inversion. Examples will illustrate the technique.

One of the elementary theorems from geometry is the statement that if a line is perpendicular to a radius at its endpoint on the circle, then it is tangent to the circle. A new theorem by inversion comes from taking the circle as the circle of inversion. It is stated as follows.

THEOREM 6.13. The circle whose diameter is the radius of the second circle is internally tangent to the second circle.

The more elementary theorem is illustrated in Figure 6.18a, whereas the inverted figure, representing Theorem 6.13, is shown in

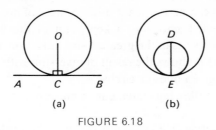

(a)   (b)

FIGURE 6.18

Figure 6.18b. It is very important to realize that the new figure is concerned only with the relationships among the image sets of points— it is not concerned with the relationship between the original figure and the new nor with areas and other extraneous factors but only with the relationship within the new figure that makes it possible to state a new theorem.

$\overline{DE}$ is the image of $\overleftrightarrow{OC}$, whereas the circle with diameter $\overline{DE}$ is the image of $\overleftrightarrow{AB}$. The circle of inversion is invariant. The fact that $\overleftrightarrow{AB}$ is tangent to the circle means that its image (a circle) is also tangent to the circle of inversion, since angles are preserved under inversion.

Figure 6.19a shows the figure for the theorem that states that an angle inscribed in a semicircle is a right angle. Figure 6.19b shows

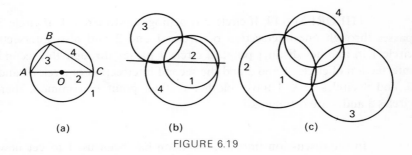

(a)   (b)   (c)

FIGURE 6.19

the result of inverting the sets of points in Figure 6.19a with respect to O as the center of inversion, with the given circle the circle of inversion. In Figure 6.19b, circle 1 and line 2 are invariant sets of points, $\overleftrightarrow{AB}$ has circle 3 as its image, and $\overleftrightarrow{BC}$ has circle 4 as its image. The new theorem involves the conclusion that circles 3 and 4 will be orthogonal. The formulation of the wording of the theorem is left as an exercise.

The choice of the center of inversion was arbitrary, and various different theorems can be written if other centers of inversion are chosen. As might be expected, choosing a key point of the figure itself as the center of inversion results in the greatest simplification, as in the previous example. Figure 6.19c shows the result of inversion with respect to a point not on any of the given lines or circles.

Verify the following statements with respect to Figure 6.19c:

1. The image of circle O, a circle not through the center of inversion, is a circle (circle 1).
2. The image of the line $\overleftrightarrow{AC}$ is a circle (circle 2) that goes through the center of inversion and is orthogonal to the image of circle O.
3. The image of the line $\overleftrightarrow{AB}$ is a circle (circle 3) that passes through the intersection of circles 1 and 2 and also through the center of inversion.
4. The image of the line $\overleftrightarrow{BC}$ is a circle (circle 4) that passes through the intersections of 1, 2 and 1, 3 and also through the center of inversion.

The new theorem will involve four circles, and the conclusion will be that two of them are orthogonal. This new theorem, valid in Euclidean geometry, is stated in somewhat complex wording, as follows:

THEOREM 6.14. If circle 2 is orthogonal to circle 1, if circle 3 passes through one intersection of circles 1 and 2 and also intersects circle 1 in another distinct point, if circle 4 passes through the second intersection of circles 1 and 2 and the second intersection of circles 1 and 3, and if circles 2, 3, 4 have another distinct point in common, then circles 3 and 4 are orthogonal.

In the discussion thus far, inversion has been used to get new theorems from old. In general, the new theorems have been more complicated because some of the straight lines have been inverted into circles. In the application of inversion called *proof by inversion*, the procedure is somewhat reversed. Here the idea is to prove a more complicated theorem by inverting the figure and then using a simpler theorem already established. The classic example of this is in the proof of Feuerbach's theorem.

THEOREM 6.15. *Feuerbach's theorem.* The nine-point circle of a triangle is tangent to the incircle and to each of the three excircles of the triangle.

In Figure 6.20, let I and I' be the centers of the inscribed circle and any one excircle, with A' the center of the side $\overline{BC}$. If

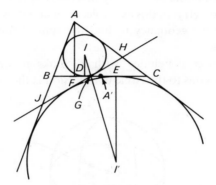

FIGURE 6.20

ID and $I'E$ are radii perpendicular to $\overline{BC}$, then A' is also the midpoint of $\overline{DE}$. This can be shown as a result of the fact that $\overline{BD} \cong \overline{CE}$, since both have measures equal to half the perimeter of the triangle minus the length of side $\overline{AC}$.

Now if A' is taken as the center of inversion and if $A'D = A'E$ is taken as the radius of inversion, then the incircle and the excircle are both invariant, since they are orthogonal to the circle of inversion. The nine-point circle passes through A' and F, the foot of the altitude from A. But A' is the center of inversion and G is the inverse of F with respect to the circle of inversion. (It can be established that G divides DE internally in the same ratio that F divides it externally.) Thus, the image of the nine-point circle is a line through G. Because angles are preserved under inversion, the angle between the image of the nine-point circle and $\overline{BC}$ is congruent to the angle that the tangent to the nine-point circle at A' makes with $\overline{BC}$. It can be shown that this tangent is parallel to the opposite side of the orthic triangle (the triangle through the feet of the altitudes). This line is antiparallel to $\overline{BC}$. Also, it can be shown that two antiparallel lines make congruent angles with the bisector of the angle between the other two sides of the inscribed quadrangle. But it is the other internal tangent of circles I and I' that makes the same angle with BC that $\overline{II'}$ does, so it is the tangent $\overleftrightarrow{HJ}$ that is the image of the nine-point circle. This means that the nine-point circle is tangent to both the incircle and the excircle. In a similar way, the nine-point circle can be shown to be tangent to the other excircles.

Another example of proof by inversion appears in Chapter 9 on non-Euclidean geometry. In this example, inversion is used in the Poincaré model for hyperbolic geometry to find the sum of the measures of the angles of a triangle.

This section concludes with a look at inversion in three dimensions. Analogous to the circle of inversion is the sphere of inversion

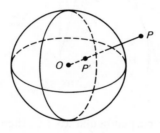

FIGURE 6.21

as shown in Figure 6.21. If P and P' are inverse points with respect to a sphere of radius r, then

$$OP \cdot OP' = r^2.$$

Inversion in two dimensions can be seen to be a cross section of the three-dimensional drawing, with the plane passing through the center of the sphere.

Corresponding to inversion relationships between the line and the circle in two dimensions are those between the plane and the sphere in three dimensions. An example is provided in the next theorem.

THEOREM 6.16. The image under inversion with respect to a sphere of a plane not through the center of inversion is a sphere passing through the center of inversion, and conversely.

A special case of inversion with respect to a sphere is known as *stereographic projection* and is very useful in map making. This method of projection is illustrated in Figure 6.22. Plane α is tangent to both the sphere with O as center and the sphere with B as the center at the common point A. Plane α and the sphere with O as center are inverses

FIGURE 6.22

with respect to the second sphere. The effect of this inversion is to establish a one-to-one correspondence between the points on the sphere and on the plane. One of the important features of a map prepared in this way is that all angle measures are preserved under inversion. Stereographic projection is also used in obtaining the Poincaré model for hyperbolic geometry mentioned above.

The inversive geometry of this chapter has proved to be a very useful geometry. Though the invariant properties are not all the same as those of Euclidean geometry, the two geometries do have many common concepts. Inversive and Euclidean geometry are distinct geometries in the sense that neither is a special case of the other. The next two geometries to be studied, projective geometry and topology, are generalizations of Euclidean geometry and include it as a special case.

EXERCISES   6.4

1. Explain the special case of Theorem 6.12 if one of the points is the center of inversion.

2. Use cardboard and paper fasteners to prepare a working model of Peucellier's cell.

3. Invert the figure of a line perpendicular to a radius of a circle at its end-point on the circle, with respect to a point on the tangent, and state the resulting theorem.

4. Invert the figure of a triangle inscribed in a semicircle with respect to one of the endpoints of the diameter and state the resulting theorem.

For Exercises 5–8, write new theorems by inverting the figure.

5. If the opposite angles of a quadrilateral are supplementary, the quadrilateral can be inscribed in a circle.

6. The line joining the centers of two intersecting circles is perpendicular to their common chord.

7. If two circles are tangent to the same line at the same point, the line joining the centers passes through the common point of tangency.

8. The altitudes of a triangle are concurrent.

Exercises 9–12 concern the proof of Theorem 6.15.

9. Verify that $\overline{BD} \cong CE$.

10. Establish that G divides $\overline{DE}$ internally in the same ratio that F divides it externally.

11. Show that the tangent to the nine-point circle at a midpoint of a side is parallel to the opposite side of the orthic triangle.

12. Prove that two antiparallel lines make congruent angles with the bisector of the angle bisector between the other two sides of the inscribed quadrangle.

13. Under inversion, with respect to a sphere, what is the image of a plane through the center of the sphere?

CHAPTER 7

PROJECTIVE GEOMETRY

7.1 FUNDAMENTAL CONCEPTS

The mathematician thinks of Euclidean geometry as a very special case of more general geometries. In this chapter, the first of these more general geometries, *projective geometry,* is explained. The first section introduces some of the basic concepts, and the second section develops the subject from an axiomatic point of view.

Special cases of projection were discussed in relation to some of the finite geometries of Chapter 1. In ordinary analytic geometry, the projection of a segment onto an axis is used. As illustrated in Figure 7.1, $\overline{AB}$ is projected onto the x-axis by dropping perpendiculars AA' and $\overline{BB'}$. Although the lengths of $\overline{AB}$ and $\overline{A'B'}$ are not the same, in general,

225

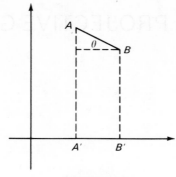

FIGURE 7.1

there is a one-to-one correspondence established by the projection between points on the two segments.

Another familiar use of the word "projection" is in connection with a motion picture projector. Here, the picture on the film is projected onto the screen, as shown in Figure 7.2a. It should be apparent that pictures on the film and on the screen will be similar to each other, since the shape remains invariant while the size changes uniformly. For

(a)   (b)

FIGURE 7.2

Figure 7.2b, however, the picture on the screen will be distorted because the film and the screen are not parallel. The figures are no longer similar, and it is no longer apparent what properties of figures remain unchanged. For example, even the ratios of distances are no longer invariant.

Although the example in Figure 7.2b may at first seem completely new, many people will be able to describe the figure as a *perspectivity,* as the term is used in art. Artists talk about centers of perspectivity to show that the lines in a painting converge at a particular point that becomes the center of attention. A famous example is the painting of the

Last Supper by Leonardo Da Vinci, shown without complete details in Figure 7.3.

FIGURE 7.3

     Famous painters during the period of the Renaissance attempted to use the concept of perspectivity to make their paintings look more realistic. They knew that the position of the artist determined how the view looked, and they attempted to put on canvas what the eye actually saw. For example, if you observe a rectangular table top from a position to one side of the table, the top no longer looks rectangular. Similarly, a circle viewed from an angle no longer looks circular. The work of the Renaissance painters can be said to have provided part of the incentive for the development of projective geometry.

     Figures 7.4a and 7.4b show sketches in which the center of

(a) (b)

FIGURE 7.4

perspectivity is not at the center. You should be able to locate this point in each case.

In this text, geometries have been studied on the basis of their invariant properties under a group of transformations. You should realize that projective geometry can also be studied using this same approach, although historically the topic was not developed that way.

It is important first to explain what is meant by a projective transformation. Figure 7.5 shows several perspectivities in the same drawing. The points on l_1 and l_2 are perspective, with A the center of

FIGURE 7.5

perspectivity. The points on l_2 and l_3 are perspective, with point B the center of perspectivity, and the points on l_3 and l_4 are perspective, with point C the center of perspectivity. In general, the points on l_1 and l_4 are not perspective. Instead, they are defined to be projective. From this point of view, a projectivity is considered a chain of perspectivities.

DEFINITION. A projectivity is a finite sequence of perspectivities.

In projective geometry, the invariant properties for a projectivity are studied. While it is not obvious which properties may be preserved, it should be evident that there is a one-to-one correspondence and that collinear points remain collinear even though distances and relative distances may change.

This intuitive discussion has introduced several of the basic concepts of projective geometry. Although it grew out of ideas developed earlier, the first real text in projective geometry, *Traité des propriétés des figures* by J. V. Poncelet (1788–1867), was published in

1822. Most of the work on this book was done while Poncelet was a prisoner in Russia, after the wars of Napoleon. Other advances in projective geometry were made during the nineteenth century. Karl von Staudt showed how projective geometry could be developed without the use of any metrical basis for measurement. Felix Klein gave projective geometry the prominent place it deserves in the classification of geometries in his Erlanger program of 1872. Since that time, postulational developments of projective geometry have been written, and finite projective geometries have been introduced.

EXERCISES 7.1

1. Write an equation relating the lengths of $\overline{AB}$ and $\overline{A'B'}$ in Figure 7.1.

2. Describe the location of the center of perspectivity in
   a. Figure 7.4a.   b. Figure 7.4b.

3. Make a rough sketch of a picture looking down a road with sets of utility poles on either side. Make the lines in the drawing converge at a point on the horizon.

For Exercises 4–11, which of these properties, which are invariant in Euclidean geometry, also seem to be invariant in projective geometry?

4. Measures of angles.

5. Collinearity.

6. Measures of area.

7. Ratios of distances.

8. Property of being a circle.

9. Property of being a square.

10. Property of being a triangle.

11. Property of being a rectangle.

12. Make a drawing similar to Figure 7.5 in which several perspectivities result in the points on a line being projected back into points on the same line.

7.2   POSTULATIONAL BASIS FOR PROJECTIVE GEOMETRY

It may seem odd at first, but the postulational basis for projective geometry is simpler from a mathematical point of view than is the basis for Euclidean geometry. The following axioms for projective geometry are adapted from those stated by H. S. M. Coxeter.[1] This set of axioms is based on the undefined terms of point, line, and *incidence*. The intuitive meaning of incidence is simply the idea of *lying on* or *containing*. For example, Axiom 1 means that the point does not lie on the line and the line does not contain the point. Axiom 2 means that every line contains at least three distinct points, or has these points on it. Axiom 3 means that two distinct points lie on just one line.

Axioms for Projective Geometry

1. There exist a point and a line not incident.
2. Every line is incident with at least three distinct points.
3. Each two distinct points are incident with exactly one line.
4. If A, B, C, D are four distinct points such that $\overleftrightarrow{AB}$ meets $\overleftrightarrow{CD}$, then $\overleftrightarrow{AC}$ meets $\overleftrightarrow{BD}$.
5. If ABC is a plane, there exists at least one point not in the plane ABC.
6. Each two distinct planes have at least two common points.
7. The three diagonal points of a complete quadrangle cannot be collinear.
8. If a projectivity leaves invariant each of three distinct points on a line, it leaves invariant every point on the line.

The vocabulary of Axiom 7 leads to new definitions.

DEFINITION. In projective geometry, a triangle consists of the three noncollinear points called vertices and the three *lines* (not segments) joining these vertices in pairs.

This definition, although not identical to that given in elementary texts, is the one often employed in higher mathematics. It is necessary in projective geometry because the concept of line segment involves the idea of betweenness, which does not appear, either explicitly or implicitly, in the axioms of projective geometry.

DEFINITION. A *complete quadrangle* is a set of four points (vertices) in a plane, no three collinear, and the lines joining these vertices in pairs.

A complete quadrangle is shown in Figure 7.6. Complete quadrangle *ABCD* has three pairs of *opposite sides*, $\overleftrightarrow{AB}$ and $\overleftrightarrow{CD}$, $\overleftrightarrow{AD}$ and $\overleftrightarrow{BC}$, and $\overleftrightarrow{AC}$ and $\overleftrightarrow{BD}$. Note that opposite sides will intersect. Opposite sides are two lines, one determined by any two vertices and the other determined by the remaining vertices of a complete quadrangle. For example, vertices *A* and *C* determine one side, so the opposite side is the line determined by the two remaining vertices, *B* and *D*.

FIGURE 7.6

The opposite sides of a complete quadrangle meet by twos in three points other than vertices (points *E, F, G* of Figure 7.6). These three points are the *diagonal points* of the complete quadrangle, as used in Axiom 7. The dotted lines in Figure 7.6 are the sides of the *diagonal triangle*, whose vertices are the diagonal points.

Many of the axioms for projective geometry seem the same as those for Euclidean geometry, and this is correct. One that is not the same, however, and that is responsible for giving projective geometry its characteristic structure is Axiom 4. This axiom is illustrated in Figure 7.7.

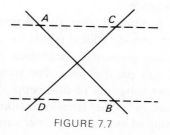

FIGURE 7.7

It is assumed in the drawing that $\overleftrightarrow{AB} \cap \overleftrightarrow{CD}$ is not empty. Axiom 4 guarantees as a result that $\overleftrightarrow{AC} \cap \overleftrightarrow{BD}$ is not empty. Since $\overleftrightarrow{AB}$ and $\overleftrightarrow{CD}$ determine a unique plane, the consequence of the axiom is that any two lines of a plane meet in a point. That is, there are no parallel lines in projective geometry. Each pair of lines in the plane determines a unique point.

While exploring the idea in the previous paragraph, it is important to understand the development leading from Euclidean geometry to projective geometry in terms of the type of elements allowed. In Euclidean geometry, every point of the Euclidean plane is an ordinary point. Two parallel lines have no point in common. In the extended Euclidean plane, parallel lines are said to meet at *ideal points*. Any two lines meet at an ordinary point or at an ideal point. Saying that two lines meet at an ideal point in the extended Euclidean plane is just another way of saying that the lines are parallel. The extended Euclidean plane consists of the union of the ordinary and ideal points in the plane. The ideal points of the extended Euclidean plane are a different concept than that of the ideal point of the inversive plane.

In the projective planes and spaces of projective geometry, the ideal points lose their special nature, and parallelism is not an invariant. It is correct to think intuitively of projective geometry as being derived from the geometry of the extended Euclidean plane through the elimination of any distinction between real and ideal points. If ideal points are not treated as special elements, the need for consideration of special cases or exceptions in much of what has been said so far in this section has been eliminated. For example, the points A, B, C, in Figure 7.5 could just as easily be ideal points as real points. In Figure 7.6, the opposite sides could intersect in ideal points or in real points. Parallelism has lost its significance, and projective geometry is displayed as a general geometry with ordinary geometry as a very special case. The axioms of projective geometry apply to points in the projective plane and in projective space.

It should be mentioned that the geometry of the extended Euclidean plane, which considers ideal points as special cases, is called *affine geometry*. In affine geometry, parallelism is preserved; parallel lines are transformed into parallel lines. The transformations of affine geometry form a proper subgroup of the group of all projective transformations. The group of affine transformations has the group of similarities and the group of motions as proper subgroups.

EXERCISES   7.2

1. Give at least one reason a person should expect the postulational basis for projective geometry to be simpler than that for Euclidean geometry.

2. Which of the eight axioms for projective geometry are also axioms for ordinary Euclidean geometry?

For Exercises 3–9, which of the eight axioms for projective geometry are also axioms for the finite geometry named?

3. Geometry of three points and three lines.

4. Geometry of four points.

5. Geometry of four lines.

6. Geometry of Fano.

7. Geometry of Pappus.

8. Geometry of Desargues.

9. Geometry of Young.

10. Prove that, in projective geometry, two distinct lines cannot have more than one point in common.

In Exercises 11 and 12, use the set of points shown in Figure 7.6.

11. If *AEGC* is considered the basic complete quadrangle, how many of its diagonal points are labeled?

12. If *AEBF* is considered the basic complete quadrangle, how many of its diagonal points are labeled?

13. Suppose the four vertices of a complete quadrangle are the vertices of a parallelogram in Euclidean geometry. What effect would this have on the diagonal points?

14. Could a triangle be a diagonal triangle for more than one complete quadrangle? Make a drawing to support your answer.

15. Each line in affine or projective geometry contains how many ideal points?

7.3   DUALITY AND SOME CONSEQUENCES

In Euclidean geometry, two points always determine exactly one line, but two lines in the same plane do not always determine a point (they may be parallel). In projective geometry, on the other hand, this exception has been eliminated. A comparison of the statements "two points determine a line" and "two lines determine a point," with respect to a plane, shows that one may be changed to the other by interchanging the words *point* and *line*. This is an example of *plane duality,* one of the interesting concepts encountered in projective geometry but not in Euclidean geometry. For sets of points in the projective plane (the plane of projective geometry), every statement remains true when the words *point* and *line* are interchanged if certain other pairs of words, such as *collinear* and *concurrent,* are changed accordingly. For example, the plane dual of the expression "three collinear points" is "three concurrent lines."

The existence and validity of the concept of plane duality in projective geometry can be shown to be a consequence of the axioms themselves. Axiom 1 is its own plane dual. The dual of Axiom 2 is a theorem that can be proved.

THEOREM 7.1. Every point is incident with at least three distinct lines.

Let $\overleftrightarrow{AB}$ and C be any line and point (Figure 7.8) not incident, as in Axiom 1. By Axiom 3, points C and A determine a second line. By Axiom 2, $\overleftrightarrow{BC}$ has a third point $D,$ and by Axiom 3, D and A determine a third line.

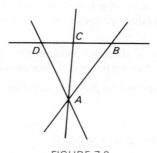

FIGURE 7.8

Axioms 3 and 4 are, in effect, duals of each other, using the interpretation of Axiom 4 as in Figure 7.7. The plane dual of Axiom 7 introduces new terminology that must be explained. This dual states that the three diagonal lines of a complete quadrilateral are never concurrent. The definition of a complete quadrilateral is written by taking the plane dual of the definition of a complete quadrangle.

DEFINITION. A *complete quadrilateral* is a set of four lines, no three concurrent, and the points of intersection of these lines in pairs (see Figure 7.9).

FIGURE 7.9

The properties of the complete quadrilateral can be studied as the plane dual of the corresponding properties of the complete quadrangle. The six points of intersection of the four given lines are in three sets of opposite vertices. In Figure 7.9, A and B are opposite vertices, as are C and D and E and F. The lines joining opposite vertices are diagonal lines, and the trilateral formed by the three diagonal lines is the *diagonal trilateral*. In projective geometry, it is customary to speak of a triangle as being self-dual, since it includes both the vertices and the sides, and to use diagonal triangle for both the complete quadrangle and the complete quadrilateral.

The dual of Axiom 7 implies that the dotted lines in Figure 7.9 cannot be concurrent. This should be easy to see, because if they were concurrent, then the three diagonal points of quadrangle $ABCD$ would be collinear, contrary to Axiom 7.

Finally, the dual of Axiom 8 states that if a projectivity leaves invariant each of three distinct lines on a point, it leaves invariant every line on the point. The assumption that this dual is a true statement results in the completion of the argument that all the duals of the axioms of projective geometry are true and that the concept of plane duality may be used freely. It is assumed that the plane dual of a definition or proved theorem is a good definition or another theorem whose proof has been established without further work.

The concept of *space duality* is based on an interchange of the words *point* and *plane,* with the word *line* being self-dual in space. In the following statements, the second gives the space dual for the first.

> Any two distinct planes have at least two common points.
>
> Space dual: Any two distinct points have at least two common planes.

One of the most fundamental, yet nontrivial, applications of the concept of plane duality is involved in the proof of Desargues' theorem and its converse. Recall that this theorem was first encountered in a finite geometry in Chapter 1.

THEOREM 7.2. *Desargues' theorem.* If two triangles are perspective from a point, they are perspective from a line.

The theorem is named for the French mathematician Desargues (1593–1662), who anticipated the development of projective geometry many years before it was actually developed. His definition (Theorem 7.2) of triangles perspective from a point and line is made clear by Figure 7.10.

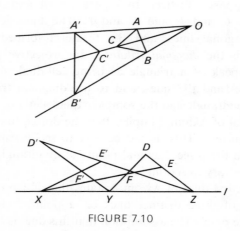

FIGURE 7.10

Two triangles such as *ABC* and *A'B'C'* are perspective from point *O*, since corresponding vertices are collinear with *O*. Two triangles such as *DEF* and $D'E'F'$ are perspective from line l, since corresponding sides meet at points X, Y, and Z on line l.

Unlike most theorems of geometry, Desargues' theorem is easier to prove for two triangles in different planes than for two triangles in the same plane.

Proof:

a. Assume that the two given triangles are in different planes, as in Figure 7.11. Since $OB'C'BC$ lie in a plane, $\overleftrightarrow{BC}$ and $\overleftrightarrow{B'C'}$ must

FIGURE 7.11

meet, and this point must be a point Z on the line of intersection of the two planes π and π'. Similarly, $OA'C'AC$ determine a plane, and $\overleftrightarrow{AC}$ and $\overleftrightarrow{A'C'}$ meet at a point Y on the line of intersection of π and π'. Lines $\overleftrightarrow{AB}$ and $\overleftrightarrow{A'B'}$ likewise must meet on the same line of intersection of π and π', so $\triangle ABC$ and $\triangle A'B'C'$ are perspective from a line.

b. Assume that the two given triangles are in the same plane. Let ABC and $A'B'C'$ be any two coplanar triangles, as in Figure 7.12a, perspective from point O. Let points C, D, E be points of intersection of pairs of corresponding sides (Figure 7.12b). We need to show that these three points are collinear.

Let O' and O'' be any two points collinear with O, but not on the plane of the given triangles (Figure 7.12c). Connect O' and the vertices of triangle $A'B'C'$ and O'' and the vertices of triangle ABC. Since the lines $O'O$ and OA determine a plane, AO'' and $O'A'$ lie in this plane and must meet at a point A''. Similarly, $O''B$ and $O'B'$ meet at B'' and $O''C$ and $O'C'$ meet at C''. Triangles ABC and $A''B''C''$ are in different planes and are perspective from O''. By Desargues' theorem for non-coplanar triangles, they are perspective from a line common to their two

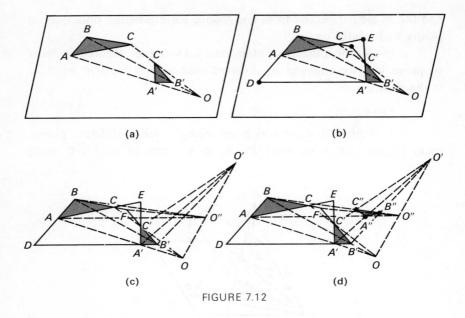

(a)   (b)   (c)   (d)

FIGURE 7.12

planes. Also, triangles $A'B'C'$ and $A''B''C''$ are noncoplanar and perspective from O', so they are also perspective from the same common line. The result is that $\overleftrightarrow{AB}$ and $\overleftrightarrow{A'B'}$ meet at a point on the common line of the plane containing the given triangles and the plane of triangle $A''B''C''$. Similarly, $\overleftrightarrow{BC}$ and $\overleftrightarrow{B'C'}$ and $\overleftrightarrow{AC}$ and $\overleftrightarrow{A'C'}$ meet on this same line and the two given triangles are perspective from this line, $\overleftrightarrow{DE}$.

From the principle of planar duality, the plane dual of Desargues' theorem has also been established.

THEOREM 7.3. If two triangles are perspective from a line, they are perspective from a point.

Note that the plane dual of Desargues' theorem is also its converse. This is not usual, but neither is it a unique case.

Desargues' theorem could be extended to a study of perspective polygons other than triangles. In such an extended study of projective geometry, you would find Desargues' theorem or its converse used frequently to help prove the concurrence of three lines or the collinearity of three points. An alternative set of axioms for projective geometry is possible by using Desargues' theorem as an axiom that eliminates the need for the space axioms.

Several other remarks should be made about Figure 7.12. The set of points and lines in the plane of the two given triangles is called the *Desargues' configuration*. It consists of ten points with three lines on each point and ten lines with three points on each line. Observe that this has already been used as the basis for a finite geometry in Chapter 1. This is an example of a finite projective geometry. In the Desargues' configuration are not only one pair of perspective triangles but a total of ten pairs, each perspective from a different point and line.

EXERCISES 7.3

Write the plane dual of each statement.

1. If *a* and *b* are distinct lines of a plane, there is at least one point on both lines.

2. On a point are at least four lines.

3. A triangle consists of three noncollinear points and the lines joining them in pairs.

Write the space dual of each of the next two statements.

4. Three planes not on the same line determine a point.

5. A plane is determined by two intersecting lines.

6. Prove the plane dual of Axiom 8.

7. Prove the converse of Desargues' theorem directly for triangles in different planes, without assuming the original theorem or using duality.

8. Draw a figure to represent Desargues' configuration, then indicate the ten pairs of perspective triangles, naming the center and axis of perspectivity for each pair.

9. Prove that if three triangles have a common center of perspectivity, their three axes of perspectivity have a common point.

10. Prove that the diagonal triangle of a quadrangle is perspective with each of the four triangles whose vertices are three of the vertices of the quadrangle.

7.4 HARMONIC SETS

Since a complete quadrangle consists of four points and six lines, an arbitrary line of the plane that does not pass through any of the four

FIGURE 7.13

vertices or any of the three diagonal points will meet the six sides in six distinct points. This set of six points is called a *quadrangular set of points*. Figure 7.13 shows complete quadrangle *ABCD* and the quadrangular set of points *E, F, G, H, I, J*. Desargues' theorem can be used to show that each point of a quadrangular set can be determined uniquely if the other five are known. The concept is introduced here, however, to give added meaning to the special case of a quadrangular set called a *harmonic set of points*.

DEFINITION. A harmonic set of points, or a *harmonic range,* is a set of four collinear points consisting of the four points of intersection of the sides of a complete quadrangle with a line passing through two diagonal points.

Figure 7.14 shows three harmonic sets of points determined by complete quadrangle *ABCD*. The four points *NFIG* are a harmonic set of points on the line passing through the diagonal points *F* and *G*. Similarly, the four points *LEMG* are a harmonic set on the line through diagonal points *E* and *G*, whereas the points *EJFK* are a harmonic set on the line through diagonal points *E* and *F*. In each case, two of the four points of the harmonic set are diagonal points, and the other two

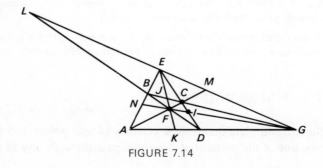

FIGURE 7.14

points are on the sides passing through the third diagonal point. A special notation for harmonic sets emphasizes this distinction between the two pairs of points in the set. For example, $H(NI, FG)$ indicates the harmonic set composed of the four points N, I and F, G. F and G are paired because they are the two diagonal points, whereas N and I are the points of intersection of the sides of the quadrangle through the third diagonal point. The same harmonic set could be indicated by $H(IN, FG)$, $H(NI, GF)$, or $H(IN, GF)$.

In the notation $H(NI, FG)$, G is the *harmonic conjugate* of F with respect to N and I. Each point of the harmonic set is the harmonic conjugate of the other member of its pair with respect to the other pair of points. Since Axiom 7 specifies that the three diagonal points of a complete quadrangle are not collinear, a point and its harmonic conjugate must be distinct points, so there are at least four points on each line in projective geometry.

The four points of a harmonic set cannot all be located independently. The statement of dependence is given in the following theorem.

THEOREM 7.4. The harmonic conjugate of a point A with respect to two other given collinear points B and C is uniquely determined.

In Figure 7.15, let B, A, C be any three given collinear points.

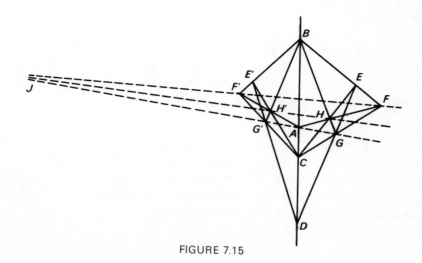

FIGURE 7.15

A harmonic conjugate D of A with respect to B and C can be found by constructing complete quadrangle $EFGH$ so that B and C are two of the diagonal points and A and D lie on lines through the third diagonal point. This can be done by choosing any point F not on $\overleftrightarrow{AB}$ and connecting F and points A, B, C. An arbitrary line can be drawn through B intersecting $\overleftrightarrow{FA}$ at H and $\overleftrightarrow{FC}$ at G. Let $\overleftrightarrow{CH}$ intersect $\overleftrightarrow{BF}$ at E. Then $\overleftrightarrow{EG}$ intersects $\overleftrightarrow{AB}$ at the required point D.

The fact that D is unique is somewhat more difficult to prove. Assume, as in Figure 7.15, that a second and distinct quadrangle $E'F'G'H'$ has been constructed as before, beginning with the choice of F' as a point not on $\overleftrightarrow{AB}$ and also distinct from F. It is required to show that $\overleftrightarrow{E'G'}$ intersects $\overleftrightarrow{AB}$ at D.

Triangles FGH and $F'G'H'$ are perspective from line $\overleftrightarrow{AB}$, hence are perspective from a point J by the converse of Desargues' theorem. Triangles FEH and $F'E'H'$ are also perspective from line $\overleftrightarrow{AB}$ and are then perspective from the same point J, so $EE'J$ are collinear. This means that triangles FEG and $F'E'G'$ are perspective from J and by Desargues' theorem are perspective from a line. This line is $\overleftrightarrow{AB}$, and $\overleftrightarrow{EG}$ and $\overleftrightarrow{E'G'}$ must meet at a point D on $\overleftrightarrow{AB}$; thus the harmonic conjugate of A with respect to B and C is a unique point.

The proof of Theorem 7.4 includes a method of constructing the harmonic conjugate of a point with respect to two other given collinear points. This construction, as anticipated in Section 5.6, uses only a straightedge.

The dual of the definition of a harmonic set of points is the definition of a harmonic set of lines.

DEFINITION. A *harmonic set of lines*, or a *harmonic pencil*, is a set of four concurrent lines such that two of them are diagonal lines of a complete quadrilateral and the other two pass through the two vertices lying on the third diagonal line.

In Figure 7.16, let a, b, c, d be the sides of the quadrilateral, with e, f, g the diagonal lines. One set of harmonic lines is $H(eg, hi)$.

The main reason for the study of harmonic sets in projective geometry is that the property of being a harmonic set is an invariant under the group of projective transformations.

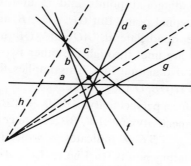

FIGURE 7.16

THEOREM 7.5. The harmonic property is preserved under projectivities.

A projectivity may be considered a finite sequence of perspectivities, so it suffices to show that the harmonic property is preserved in a single perspectivity. To prove that this is so consists of proving two separate statements:

a. The set of lines joining any noncollinear point to the four points of a harmonic set of points is a harmonic set of lines.

b. The set of points of intersection of the four lines of a harmonic set with any line not through the point of concurrency of the lines (the center of the pencil) is a harmonic set of points.

Since the two statements are plane duals, proving one will be sufficient. The first statement is proved here. In Figure 7.17, let

FIGURE 7.17

$H(AB, CD)$ be any harmonic set of points with O any noncollinear point. This harmonic set implies the existence of a complete quadrangle $OEFG$

with A and B two diagonal points and C and D points on lines through the third diagonal point. But $\overleftrightarrow{GF}$, $\overleftrightarrow{AE}$, $\overleftrightarrow{AB}$, and $\overleftrightarrow{GE}$ are the four sides of a complete quadrilateral. $\overleftrightarrow{AO}$ and $\overleftrightarrow{OB}$ are diagonal lines, whereas $\overleftrightarrow{OC}$ and $\overleftrightarrow{OD}$ are lines through the other two vertices, F and D, which lie on the third diagonal line. This implies the existence of the harmonic set of lines $H(\overleftrightarrow{OB}, \overleftrightarrow{OA}, \overleftrightarrow{OC}, \overleftrightarrow{OD})$.

The symbol for a perspectivity, $ABCD \overset{S}{\overline{\wedge}} A'B'C'D'$, indicates that S is the center of perspectivity and AA', BB',... are corresponding points. The symbol $ABCD \overline{\wedge} A'B'C'D'$ indicates a projectivity in which AA', BB',... are corresponding points. In Figure 7.17, $ABCD \overset{O}{\overline{\wedge}} GEHD$ and $GEHD \overset{I}{\overline{\wedge}} BACD$, so $ABCD \overline{\wedge} BACD$. This implies that:

THEOREM 7.6. If $H(AB, CD)$, then $H(BA, CD)$.

Also in Figure 7.17, $ABCD \overset{E}{\overline{\wedge}} FOHC$ and $FOHC \overset{G}{\overline{\wedge}} BADC$, so that:

THEOREM 7.7. If $H(AB, CD)$, then $H(BA, DC)$.

Beginning with three distinct points on a line, the harmonic conjugate of one of these points with respect to the other two can be uniquely determined. For the set of four points, each set of three determines other harmonic conjugates. All the harmonic conjugates on a line determined by such a sequence of steps are said to be *harmonically related* to the three original points.

DEFINITION. The set of all points harmonically related to three distinct collinear points is a *harmonic net* of points.

The concept of harmonic net is fundamental in the development of a system of coordinates for points in projective geometry (see Section 7.6).

EXERCISES   7.4

Given three points A, B, C in this order on a line:

1. Locate D by construction with a straightedge so that $H(AC, BD)$ is a harmonic set.

2. Locate D by construction with a straightedge so that $H(AB, CD)$ is a harmonic set.

3. Use Theorems 7.6 and 7.7 to write all possible pairings obtained from $H(CX, DY)$.

4. Suppose, in the Euclidean sense, that B is the midpoint of $\overline{AC}$. Describe the location of the harmonic conjugate of B with respect to A and C.

For Exercises 5–9, which of these are harmonic sets in Figure 7.14? Why?

5. (AD, KG) 6. (AC, FM) 7. (AB, NE) 8. (LF, BD) 9. (BC, JG)

10. Prove statement b under Theorem 7.5 directly without using duality.

11. Prove that the sixth point of a quadrangular set of points can be determined uniquely if the other five are known.

12. Prove that a line in projective geometry has more than four distinct points.

7.5 PROJECTIVITIES

A projectivity has been presented as a finite sequence of perspectivities. A projectivity is a type of transformation, since it is a one-to-one onto mapping of points.

The inverse of a projectivity is a projectivity, and the product of two projectivities is a projectivity; thus, the following theorem is true (Exercise 9, Exercise Set 7.5).

THEOREM 7.8. The set of all projectivities of the plane constitutes a group of transformations.

Projective geometry is the study of properties of sets of points that are invariant under the group of projective transformations. The algebraic approach to the study of projective geometry must wait until the introduction of coordinates for the projective plane in the next section. Before this, however, it is possible to study projectivities from a different viewpoint. A question that may have occurred to you is: What is the minimum information needed to uniquely determine a projectivity? Since projectivities involve points and their images, how many pairs of points and images must be given so that all pairs can be determined? The answers to these questions appear in what is called *the fundamental theorem of projective geometry*.

THEOREM 7.9. *Fundamental theorem.* A projectivity between the sets of points on two lines in a plane is determined by three collinear points and their images.

The proof is indirect. Assume, as illustrated in Figure 7.18, that $ABCX \barwedge A'B'C'X'$, and $ABCX \barwedge A'B'C'X''$, for X' and X'' distinct points.

FIGURE 7.18

Then $A'B'C'X' \barwedge ABCX \barwedge A'B'C'X''$, and $A'B'C'X' \barwedge A'B'C'X''$. But in this last projectivity, points A', B', and C' are fixed. By Axiom 8 of Section 7.2, all the other points on the line must also be fixed. The assumption that X' and X'' are distinct contradicts the axiom, so X' and X'' must be identical.

While the fundamental theorem of projective geometry is very important from a theoretical point of view, it does not yet provide a constructive method for actually determining additional pairs of corresponding points in a projectivity. One such method is based on establishing a minimum sequence of perspectivities for the projectivity. This sequence, while not unique, yields unique results because of the fundamental theorem.

THEOREM 7.10. Three distinct points A, B, C on one line may be projected into any three distinct points A', B', C' on a second line by means of a sequence of at most two perspectivities.

In Figure 7.19, let A, B, C, A', B', C' represent the six given points. Choose an arbitrary center of perspectivity on $\overleftrightarrow{AA'}$ and an arbitrary distinct line l through A'. Then points B'' and C'' are determined on l so that $ABC \mathbin{\overset{S}{\wedge}} A'B''C''$. The intersection of $\overleftrightarrow{C'C''}$ and $\overleftrightarrow{B'B''}$ determines a point S' such that $A'B''C'' \mathbin{\overset{S'}{\wedge}} A'B'C'$, so $ABC \mathbin{\overset{}{\wedge}} A'B'C'$.

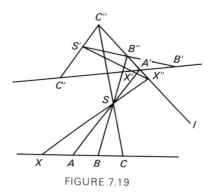

FIGURE 7.19

Given any point X on $\overleftrightarrow{AB}$, X'' and X' can be found by using the perspectivities already set up. If the six given points all lie on the same line, then one extra perspectivity is required. See Exercise 4 of Exercise Set 7.4.

The corresponding points in a projectivity are not, in general, perspective from a point, nor are the corresponding lines in a projectivity perspective from a line. But projectivities do have a point or a line that is somewhat analogous to centers and axes of perspectivity. These concepts are useful in Section 7.8.

THEOREM 7.11. A projectivity between two sets of points on two distinct lines determines a third line called the *axis,* or *axis of homology,* and contains the intersections of the cross joins of all pairs of corresponding points.

DEFINITION. The *cross joins* of two pairs of points, such as A, A' and B, B', are $\overleftrightarrow{AB'}$ and $\overleftrightarrow{BA'}$.

To prove Theorem 7.11, it must be shown that the axis is unique and that all of the intersections of the cross joins are collinear. This can be accomplished by showing that the axis is independent of the choices of points.

Let D, Figure 7.20, be the point of intersection of the two given lines. Let E be the image of D when D is considered as a point on $\overleftrightarrow{DA'}$, and let E' be the image of D considered as a point on $\overleftrightarrow{DA}$.

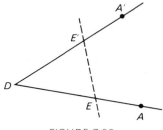

FIGURE 7.20

For the two pairs of points A, A' and D, E', the cross joins are $\overleftrightarrow{AE'}$ and $\overleftrightarrow{A'D}$, which meet at point E' on the axis of homology. For the two pairs of points A', A and D, E, the cross joins are $\overleftrightarrow{A'E}$ and $\overleftrightarrow{AD}$, meeting at E. Thus, the axis of homology always passes through the two images of the common point of the two lines, when the common point is first considered as an element of one line and then of another. The axis of homology is thus seen to be independent of the pairs of points selected.

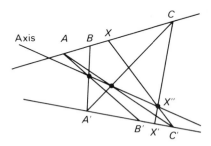

FIGURE 7.21

The axis of homology may be used to construct additional pairs of corresponding points in a projectivity, as shown in the following example. In Figure 7.21, assume that a projectivity is determined by the given pairs of points A, A', B, B', C, C'. The axis of homology can be determined by the intersection of the cross joins $\overleftrightarrow{AC'}, \overleftrightarrow{A'C}$ and $\overleftrightarrow{AB'}, \overleftrightarrow{A'B}$. Let X represent any fourth point on $\overleftrightarrow{AB}$. Now $\overleftrightarrow{XC'}$ intersects the axis of homology at a point X'', and $\overleftrightarrow{CX''}$ meets $\overleftrightarrow{A'C'}$ at the required point X' on $\overleftrightarrow{A'C'}$. Note that this construction uses only a straightedge.

The duals of the theorems already presented in this section are interesting in their own right. The proofs given depend on the concept of duality, even though the theorems could be proved directly.

THEOREM 7.12. *Dual of the fundamental theorem.* A projectivity between the sets of lines on two points in a plane is determined by three concurrent lines and their images.

THEOREM 7.13. *Dual of Theorem 7.11.* A projectivity between two sets of lines on two distinct points determines a third point, called the *center*, or *center of homology*, which lies on the joins of the cross intersections of corresponding lines.

Theorem 7.13 is illustrated in Figure 7.22. Lines *a, b, c* and

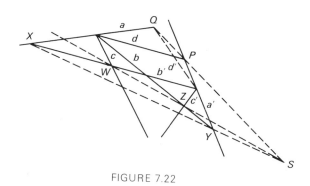

FIGURE 7.22

a', b', c' are given. Point X is the intersection of a and b' and Y is the intersection of a' and b, so $\overleftrightarrow{XY}$ passes through S, the center of homology. Likewise, c and b' intersect at W and c' and b meet at Z, so S lies on $\overleftrightarrow{WZ}$.

The center of homology can be used to construct additional

pairs of corresponding lines in a projectivity. In Figure 7.22, suppose that d is given. Lines d and a' meet at P. Line a meets d' at point Q on $\overleftrightarrow{PS}$. The line connecting Q and the point of concurrency of a', b', c' is the desired line d'.

EXERCISES 7.5

For Exercises 1–3, use Figure 7.23.

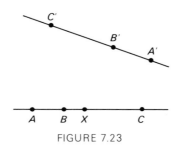

FIGURE 7.23

1. Use the method in the proof of Theorem 7.10 to establish two perspectivities so that a projectivity is established between A, B, C and A', B', C'.

2. Construct the axis of homology for the projectivity with A, A', B, B', and C, C' as corresponding points.

3. Find the image of point X in the projectivity of Exercise 2.

4. Prove that three distinct points A, B, C on a line can be projected into any three distinct points A', B', C' on the same line by means of a chain of three perspectivities.

5. Copy Figure 7.24. Set up a chain of three perspectivities so that A and A', B and B', and C and C' correspond.

FIGURE 7.24

6. For the projectivity in Exercise 5, find the image of X.

7. Copy Figure 7.25. Find the point of homology for the projectivity with aa', bb', and cc' as corresponding lines.

8. Find the image of line X in the projectivity of Exercise 7.

9. Complete the details of the proof of Theorem 7.8.

10. Prove the plane dual of the fundamental theorem, without using duality.

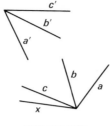

FIGURE 7.25

7.6 HOMOGENEOUS COORDINATES

This section presents a brief analytic treatment of some of the fundamental ideas of projective geometry. Since the projective plane includes ideal points that are no longer considered as special elements, the Cartesian coordinate system of ordinary analytic geometry will no longer suffice.

Prior to a more formal presentation of the coordinate system of projective geometry, called *homogeneous coordinates,* it is important to see intuitively how the two systems of coordinates are related. Figure 7.26a

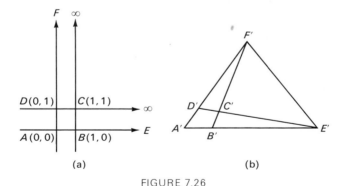

(a)   (b)

FIGURE 7.26

shows the framework of the ordinary Cartesian coordinate system, with points E and F indicated as "points at infinity" in a coordinate system for affine geometry. Figure 7.26a may be projected into Figure 7.26b, with the corresponding elements indicated by primes. The ideal points E' and F' are no longer special points, and $\overleftrightarrow{E'F'}$ is the ideal line. The points in the first quadrant of Figure 7.26a are projected into the points in the interior of triangle $A'E'F'$. Already, it should be apparent that the coordinate system for projective geometry is not one in which distance is preserved in the usual sense.

The homogeneous coordinates must be defined in such a way that they will distinguish among an infinite number of ideal points and yet not require any restrictions in the way these points are handled. The substitution

$$x = \frac{x_1}{x_3}, \qquad y = \frac{x_2}{x_3} \quad \text{for} \quad x_3 \neq 0$$

relates the ordinary Euclidean point with coordinates (x, y) to the corresponding point in the projective plane with homogeneous coordinates (x_1, x_2, x_3). For example, $(2, 3)$ becomes $(2, 3, 1)$; $(3, 4, 5)$ is the same point as $(3/5, 4/5)$. For ideal points, the restriction $x_3 \neq 0$ must be removed, and the stage is set for a more formal approach.

DEFINITION. In analytic projective geometry, a *point* is an ordered triple (x_1, x_2, x_3), not all zero.

The notation (ax_1, ax_2, ax_3) represents the same point as (x_1, x_2, x_3) for any nonzero a. For example, $(3, 1, 2)$ and $(6, 2, 4)$ name the same point in homogeneous coordinates. A line in homogeneous coordinates is defined dually as an ordered triple $[X_1, X_2, X_3]$, with the understanding that $[aX_1, aX_2, aX_3]$ names the same line for any nonzero a. For example,

$$[3, 2, 7] \quad \text{and} \quad [1, \tfrac{2}{3}, \tfrac{7}{3}]$$

are two names for the same line.

A point x lies on a line X if and only if

$$X_1 x_1 + X_2 x_2 + X_3 x_3 = 0.$$

For example, the point $(2, 3, 0)$ lies on the line $[3, -2, 0]$, since $(2 \cdot 3) + (3 \cdot -2) + (0 \cdot 0) = 0$.

The condition in the previous paragraph is a generalization of the condition for points lying on lines in Euclidean analytic geometry. If x_3 and X_3 are set equal to 1, the condition reduces to $X_1 x_1 + X_2 x_2 + (1 \cdot 1) = 0$, which is equivalent to the condition

$$\left(\frac{a}{c}\right)x_1 + \left(\frac{b}{c}\right)y_1 + (1 \cdot 1) = 0$$

for the point (x_1, y_1) to lie on the line

$$\frac{a}{c}x + \frac{b}{c}y + 1 = 0.$$

For a variable point x on the line X, the equation of the line is given by $X_1 x_1 + X_2 x_2 + X_3 x_3 = 0$. For example, the equation of the line $[2, -5, 7]$ is $2x_1 - 5x_2 + 7x_3 = 0$. Dually, the equation of the point $(3, 1, 2)$ is $3X_1 + X_2 + 2X_3 = 0$. Points do not have equations in Euclidean analytic geometry, but the duality in the projective plane should have prepared you for their appearance in the study of homogeneous coordinates.

It is now possible to relate the formal definitions of homogeneous coordinates to Figure 7.26, recalling the intuitive development in terms of ratios. Study Figure 7.27 carefully, noting the homogeneous coordinates of each point as related to the ordinary coordinates of the corresponding points.

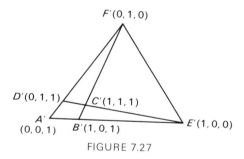

FIGURE 7.27

From Figure 7.27, it is not difficult to determine the equations of the various lines and to relate them to the corresponding lines in Euclidean geometry. The relationships are as follows.

$$\overleftrightarrow{A'E'} \qquad x_2 = 0$$

$$\overleftrightarrow{A'F'} \qquad x_1 = 0$$

$$\overleftrightarrow{E'F'} \qquad x_3 = 0$$

$$\overleftrightarrow{B'C'} \qquad x_1 - x_3 = 0$$

$$\overleftrightarrow{D'C'} \qquad x_2 - x_3 = 0$$

These equations can be derived formally from the condition $X_1 x_1 + X_2 x_2 + X_3 x_3 = 0$ for a point to lie on a line, by substituting the coordinates of the two points on a line and then solving the set of simultaneous equations. For example, for $\overleftrightarrow{C'D'}$,

$$X_2 + X_3 = 0,$$
$$X_1 + X_2 + X_3 = 0.$$

Let $X_3 = 1$; therefore,

$$X_2 = -1,$$
$$X_1 = 0,$$

and the equation is $-x_2 + x_3 = 0$ or $x_2 - x_3 = 0$. For the lines given, the equations can also be derived intuitively simply by examining the coordinates to note a pattern.

It should be understood that the triangle in Figure 7.27 includes not the entire projective plane but only that part corresponding to the first quadrant. The points $A'E'F'$ are vertices of what is called the *fundamental triangle*, whereas C', with coordinates $(1, 1, 1)$, is called the *unit point*.

The procedure by which coordinates are found for additional points in the projective plane is based upon the concept of a harmonic net. This procedure is illustrated for the line $x_2 = 0$ in Figure 7.28.

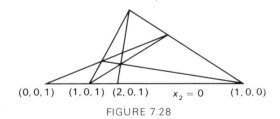

(0,0,1) (1,0,1) (2,0,1) $x_2 = 0$ (1,0,0)

FIGURE 7.28

The point $(2, 0, 1)$ is defined as the harmonic conjugate of $(0, 0, 1)$ with respect to $(1, 0, 1)$ and $(1, 0, 0)$. The point can be constructed with a straightedge as explained in Section 7.4. The series of constructions could be carried on to show that $(3, 0, 1)$ is the harmonic conjugate of $(1, 0, 1)$ with respect to $(2, 0, 1)$ and $(1, 0, 0)$. Also, $(\frac{1}{2}, 0, 1)$ is the harmonic conjugate of $(1, 0, 0)$ with respect to $(0, 0, 1)$ and $(1, 0, 1)$. The procedure can be continued for points with negative coordinates. For example, $(-1, 0, 1)$ is the harmonic conjugate of $(1, 0, 1)$ with respect to $(0, 0, 1)$ and $(1, 0, 0)$. Not all the points on the line, but all those in the harmonic net determined by the three beginning points, can be constructed.

EXERCISES 7.6

1. How can one tell an ideal point by looking at its coordinates?
2. Write homogeneous coordinates for these points:
   a. $(3, 8)$ b. $(2, \frac{3}{4})$ c. $(1, -4)$
3. Write Cartesian coordinates for these points:
   a. $(2, 5, 1)$ b. $(-1, 3, 2)$ c. $(-2, 5, -3)$
4. Write three other names, using homogeneous coordinates, for the point $(5, 2, 3)$.
5. Does the point $(1, 1, 1)$ lie on the line $[1, 1, 1]$?
6. Find the coordinates for the ideal point on the line $[1, 1, 1]$.
7. Write the equation of the line $[3, 1, -2]$.
8. Write the equation of the point $(3, 1, -2)$.
9. Develop the equation for $\overleftrightarrow{B'C'}$, Figure 7.27.
10. Find the point of intersection of the lines $[1, 1, 1]$ and $[2, 1, 2]$.
11. Find the equation of the line through $(1, 1, 1)$ and $(2, 1, 2)$.
12. Give the coordinates of the harmonic conjugate of $(0, 1, 0)$ with respect to $(0, 1, 1)$ and $(0, 3, 1)$.

For Exercises 13–16, construct with a straightedge the required points on the line $x_2 = 0$, given $(0, 0, 1)$, $(1, 0, 1)$, and $(1, 0, 0)$.

13. $(2, 0, 1)$ 14. $(3, 0, 1)$
15. $(-1, 0, 1)$ 16. $(\frac{1}{2}, 0, 1)$

7.7 EQUATIONS FOR PROJECTIVE TRANSFORMATIONS

The set of simultaneous equations needed to express a projective transformation analytically is more difficult to develop than were the sets of equations for transformations given so far—those for Euclidean motions, for example. Keep in mind the fact that, in projective geometry, straight lines are transformed into straight lines so the set of equations for projective transformations is a set of three simultaneous linear equations relating the homogeneous coordinates of a point to the homogeneous coordinates of its image.

Three points (x_1, x_2, x_3), (y_1, y_2, y_3), (z_1, z_2, z_3) are collinear if and only if there exist three numbers a, b, c, not all zero, such that $ax_i + by_i + cz_i = 0$. This means that any point on a line is a linear combination of any other two distinct points on the line. For example, the point $(8, 13, 23)$ is on the line with points $(1, 2, 4)$ and $(2, 3, 5)$, since

$$(2)(1) + (3)(2) + (-1)(8) = 0,$$
$$(2)(2) + (3)(3) + (-1)(13) = 0,$$
$$(2)(4) + (3)(5) + (-1)(23) = 0.$$

In this example, $a = 2$, $b = 3$, and $c = -1$.

No three of the four basic points $(1, 0, 0)$, $(0, 1, 0)$, $(0, 0, 1)$, $(1, 1, 1)$ are collinear. These four points may be transformed by a projective transformation into four noncollinear points d, e, f, and $(d + e + f)$ whose coordinates are

$$(d_1, d_2, d_3),$$
$$(e_1, e_2, e_3),$$
$$(f_1, f_2, f_3),$$

and

$$(d_1 + e_1 + f_1, d_2 + e_2 + f_2, d_3 + e_3 + f_3).$$

This transformation can be accomplished by the correspondence below:

$$x_1' = d_1 x_1 + e_1 x_2 + f_1 x_3,$$
$$x_2' = d_2 x_1 + e_2 x_2 + f_2 x_3,$$
$$x_3' = d_3 x_1 + e_3 x_2 + f_3 x_3.$$

This set of equations transforms any other point (k, l, m) into a point whose coordinates are given by the equations:

$$x_1' = d_1 k + e_1 l + f_1 m,$$
$$x_2' = d_2 k + e_2 l + f_2 m,$$
$$x_3' = d_3 k + e_3 l + f_3 m.$$

The new point may be indicated by $(dk + el + fm)$, which shows that it has corresponding coordinates referred to points d, e, f rather than the original vertices of the fundamental triangle, since (k, l, m) could have been indicated by the coefficients in the expression

$$([k \cdot 1] + [l \cdot 1] + [m \cdot 1]).$$

The set of linear equations given are the simultaneous equations sought to represent the projective transformation. They provide a one-to-one correspondence of points in the projective plane. They preserve collinearity of points. Although the intuitive development has not been complete, the theorem is stated, with additional discussion following.

THEOREM 7.14. A set of equations representing a projective transformation is of the form

$$x_1' = d_1 x_1 + e_1 x_2 + f_1 x_3,$$
$$x_2' = d_2 x_1 + e_2 x_2 + f_2 x_3,$$
$$x_3' = d_3 x_1 + e_3 x_2 + f_3 x_3.$$

It is understood that the determinant of the coefficients is not zero. Also, because homogeneous coordinates are not unique but may be multiplied by a constant and still name the same number, the introduction of a parameter as a coefficient of the primed terms is considered carefully in a more detailed course.

The proof of one and the assumption of two additional theorems will help establish that the equations of Theorem 7.14 do provide for a projective transformation because they keep projective properties invariant.

THEOREM 7.15. A transformation in the plane is a projective transformation if it transforms one line projectively.

In Figure 7.29, let l and l' be the given lines with the projective transformation and m and m' any other pair of corresponding lines. Let X and X' represent corresponding points on l and l', with A and B any other two distinct points not on any of the four lines. Line

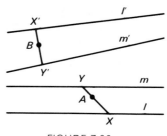

FIGURE 7.29

XAY is transformed into $X'BY'$, and since Y and Y' are projectively related, the following chain shows the projectivity connecting l and l'.

$$X \overset{A}{\barwedge} Y \wedge Y' \overset{B}{\barwedge} X'$$

The use of Theorem 7.15 means that the homogeneous co-ordinates may be considered for one dimension only. Projective co-ordinates for three points on a line can be chosen so that they are (x_1, x_2), (y_1, y_2), and $(x_1 + y_1, x_2 + y_2)$. Then the harmonic conjugate of the third point with respect to the first two has the coordinates $(x_1 - y_1, x_2 - y_2)$. For example, the harmonic conjugate of $(4, 3)$ with respect to $(1, 1)$ and $(3, 2)$ is $(-2, -1)$.

A generalization of the concept of harmonic set is the idea of the *cross ratio* of four points.

DEFINITION. The cross ratio r of four collinear points is the number r when the coordinates of the four points are written in the form (x_1, x_2), (y_1, y_2), $(x_1 + y_1, x_2 + y_2)$, $(rx_1 + y_1, rx_2 + y_2)$.

For example, the cross ratio for $A(2, 3)$, $B(5, 6)$, $C(7, 9)$, $D(9, 12)$, with the pairing as in the definition, is 2, since $9 = (2 \cdot 2) + 5$ and $12 = (2 \cdot 3) + 6$.

Two additional theorems, stated without proof, will make a conclusion possible.

THEOREM 7.16. The cross ratio of four points is an invariant under projection.

THEOREM 7.17. A transformation that preserves the cross ratio of every four collinear points is a projective transformation.

Theorem 7.16 and Theorem 7.17 may be used for the line $x_1 = 0$. The set of simultaneous equations given in Theorem 7.14 transforms a point, $(0, x_2, x_3)$ on $x_1 = 0$, into $(e_1 x_2 + f_1 x_3, e_2 x_2 + f_2 x_3, e_3 x_2 + f_3 x_3)$. The four points $(0, 0, 1)$, $(0, 1, 1)$, $(0, 1, 0)$, and $(0, x_2, x_3)$ have a cross ratio of x_2/x_3, since

$$(0, x_2, x_3) = \left(0, \frac{x_2}{x_3}, 1\right)$$

It can be verified that these four points are projected into four points with this same cross ratio. Since the four sets of coordinates chosen may represent any four points on the line, the line is transformed projectively.

It has now been established that every projectivity in the projective plane has equations of the form

$$x_1' = a_1 x_1 + a_2 x_2 + a_3 x_3,$$

$$x_2' = b_1 x_1 + b_2 x_2 + b_3 x_3,$$

$$x_3' = c_1 x_1 + c_2 x_2 + c_3 x_3,$$

with the stipulation that the determinant of the coefficients,

$$\begin{vmatrix} a_1 & a_2 & a_3 \\ b_1 & b_2 & b_3 \\ c_1 & c_2 & c_3 \end{vmatrix}$$

is not zero. In linear algebra, the concept of a *matrix*, a rectangular array of numbers, is utilized to analyze sets of linear equations. The set of equations for the projective plane has the matrix of coefficients

$$\begin{pmatrix} a_1 & a_2 & a_3 \\ b_1 & b_2 & b_3 \\ c_1 & c_2 & c_3 \end{pmatrix}$$

This convenient notation makes it easy to symbolize various subgroups of the set of projective transformations previously introduced by writing the matrix of the coefficients. This is illustrated by the following matrices, with the determinant nonzero in all cases.

$$\begin{pmatrix} a_1 & a_2 & a_3 \\ b_1 & b_2 & b_3 \\ 0 & 0 & c_3 \end{pmatrix}$$ affine transformations

$$\begin{pmatrix} a_1 & a_2 & a_3 \\ \pm(-a_2) & (\pm a_1) & b_3 \\ 0 & 0 & 1 \end{pmatrix}$$ similarity transformations
$(a_1^2 + a_2^2 \neq 0)$

$$\begin{pmatrix} a_1 & a_2 & a_3 \\ \pm(-a_2) & (\pm a_1) & b_3 \\ 0 & 0 & 1 \end{pmatrix}$$ motions
$(a_1^2 + a_2^2 = 1)$

The uses of matrices in connection with transformations can be greatly extended in an appropriate course. For example, the definition of matrix multiplication can be used to find the product of transformations, and the concept of inverse of a matrix can be used to find the inverse of a transformation.

EXERCISES   7.7

1. Find numbers a, b, c to show that the line through $(1, 1, 1)$ and $(3, 2, 1)$ also passes through $(7, 4, 1)$.

For Exercises 2 and 3, find the harmonic conjugate of the third point with respect to the first two.

2. $(3, 1)$, $(-2, 5)$, $(1, 6)$

3. $(-1, -2)$, $(3, 2)$, $(2, 0)$

Use the projective transformation

$$x_1' = x_1 + x_2 + x_3,$$
$$x_2' = x_1 - x_2 + x_3,$$
$$x_3' = x_1 + x_2 - x_3$$

for Exercises 4–6.

4. Find the images of the points $(0, 0, 1)$, $(1, 0, 0)$, $(1, 0, 1)$, and $(3, 0, 5)$.

5. Find the cross ratio of the four original points in Exercise 4.

6. Verify that the cross ratio of the images is the same as the original cross ratio in Exercise 5.

For Exercises 7–10, what type of transformation, if any, is represented by each matrix of coefficients?

7. $\begin{pmatrix} 2 & 3 & 4 \\ -3 & 2 & 1 \\ 0 & 0 & 1 \end{pmatrix}$

8. $\begin{pmatrix} 1 & 2 & 3 \\ 4 & 5 & 6 \\ 7 & 8 & 9 \end{pmatrix}$

9. $\begin{pmatrix} 1 & 2 & 3 \\ 4 & 5 & 6 \\ 0 & 0 & 5 \end{pmatrix}$

10. $\begin{pmatrix} 1 & 0 & 2 \\ 0 & 1 & 3 \\ 0 & 0 & 1 \end{pmatrix}$

7.8   SPECIAL PROJECTIVITIES

The general theory of projectivities developed in Section 7.5 can be both used and extended by considering special projectivities in one and two dimensions.

A one-dimensional projectivity renames the points on the same line.

DEFINITION. A one-dimensional projectivity is called *elliptic, parabolic,* or *hyperbolic* if the number of invariant points is zero, one, or two, respectively. If there are three invariant points, the projectivity is the identity transformation.

The fundamental theorem of projective geometry can be used to state that:

1. A hyperbolic projectivity is determined when both invariant points and one other set of corresponding points are given.
2. A parabolic projectivity is determined when its invariant point and two other sets of corresponding points are given.

This second statement can be sharpened significantly, however, by proving the following theorem:

THEOREM 7.18. A parabolic projectivity is determined when its invariant point and one other set of corresponding points are given.

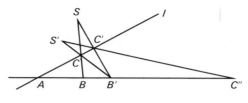

FIGURE 7.30

In Figure 7.30, let a parabolic projectivity be given, with A the invariant point and B and B' any other pair of corresponding points. An arbitrary point S can be chosen so that $ABB' \overset{S}{\wedge} ACC'$, where A, C, C' are on a line l through A not containing S. A second center of perspectivity S' can be chosen so that $ACC' \overset{S'}{\wedge} AB'C''$. The image of any other point D on $\overleftrightarrow{AB}$ can be found by using the two perspectivities established. From what has been said so far, it has not been proved that the projectivity established is indeed parabolic rather than hyperbolic. That is, for some point D on line AB, the two perspectivities might result in D as the image, giving a second invariant point. For this to happen, D, S, and S' must be collinear, a situation that can be avoided simply by inserting the restriction that S and S' lie on the same line through A.

Some special projectivities are periodic in the sense that they will result in the identity transformation after a finite number of repeated applications.

DEFINITION. A projectivity of *period n* is one that must be repeated n times before first resulting in the identity transformation.

DEFINITION. An *involution* is a projectivity of period two.

It can be seen intuitively that a one-dimensional involution simply interchanges pairs of points. The following theorem establishes a minimum condition for a projectivity to be an involution.

THEOREM 7.19. A one-dimensional projectivity that exchanges one pair of distinct points is an involution.

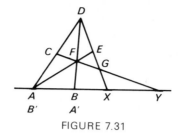

FIGURE 7.31

In Figure 7.31, let ABX be any three given distinct collinear points, with $A'B'Y$ their images in a projectivity that interchanges A and B. These six points determine a unique projectivity that can be represented by these perspectivities: $ABXY \overset{F}{=} EDXG \overset{A}{\overline{\wedge}} FCYG \overset{D}{\overline{\wedge}} BAYX$, so $ABXY \overline{\wedge} BAYX$. This projectivity interchanges all arbitrary pairs of points XY and is an involution.

Some special one-dimensional transformations have been mentioned, and one-dimensional transformations are still involved in the discussion of the projective geometry of the plane. A two-dimensional projectivity transforms every one-dimensional set projectively. That is, a two-dimensional projective transformation may involve every point of the plane, but each line in the plane is transformed into another line so that there is a projectivity established by the points on the two lines.

THEOREM 7.20. A two-dimensional projectivity that leaves the four lines of a complete quadrilateral invariant is the identity transformation.

As illustrated in Figure 7.32 for quadrilateral a, b, c, d, the six

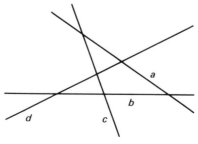

FIGURE 7.32

vertices lie three on each line, and these vertices are invariant points. There is a projectivity between two corresponding sides, and every point on these sides must be invariant since three pairs are. Any other line meets the sides of this quadrilateral in invariant points and every point on it must be invariant, so the transformation is the identity.

Special types of two-dimensional projectivities include one relating two given perspective triangles, called a *perspective collineation*. That is, the two triangles are images under a projectivity and are also perspective. The point and line of perspectivity are the *center* and *axis* of the transformation. The special case for which the center lies on the axis is called an *elation*, while all other transformations of this type are *homologies*. A special case is a homology in which the harmonic conjugate of the center with respect to pairs of corresponding points is on the axis. This transformation is called a *harmonic homology* and is illustrated by the example in Figure 7.33. Triangles ABC and $A'B'C'$ are perspective from point S and from line l. Furthermore, the harmonic conjugate of S with respect to pairs of corresponding vertices is on line l. For example, S' is the harmonic conjugate of S with respect to A and A'.

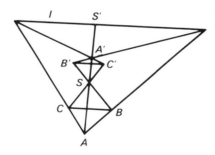

FIGURE 7.33

If the center and axis are given, the pairs of corresponding points in a harmonic homology can be determined, since they are harmonic conjugates with respect to the center and a collinear point on the axis. This type of transformation is of period two, since repeating it will result in the identity. In fact, it can be established that every two-dimensional transformation of period two is a harmonic homology.

EXERCISES 7.8

1. Explain why a reflection is an example of an involution.

2. Give an example of a rotation that is of period

    a. Two.   b. Three.   c. Four.

3. Show how the image of any other point on the line can be constructed if the invariant point and one other set of corresponding points are given for a one-dimensional parabolic projectivity.

4. Show how the image of any other point on the line can be constructed if the invariant points and one other set of corresponding points are given for a one-dimensional hyperbolic projectivity.

5. State and prove a theorem about the information needed to determine a unique one-dimensional involution.

6. State the plane dual of Theorem 7.20 and prove it by writing the dual of the proof of Theorem 7.20.

7. Sketch an example of an elation.

8. Sketch a homology that is not a harmonic homology.

9. In Figure 7.34, let S and l determine a harmonic homology. Construct the images of points A, B, C.

FIGURE 7.34

7.9 CONICS

You should recall from elementary geometry that the various conics of Euclidean geometry can all be described as sections of cones of two nappes, as illustrated in Figure 7.35. From a different point of view, Figure 7.35 shows that a circle can be projected into an ellipse (Figure 7.35a), a parabola (Figure 7.35b), or a hyperbola (Figure 7.35c). The property of being a conic is an invariant under the group of projective transformations, but the property of being an ellipse, a parabola, or a hyperbola is not.

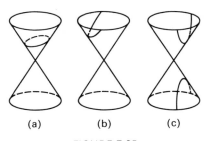

(a)   (b)   (c)

FIGURE 7.35

Just as there is more than one way to define conics in Euclidean analytic geometry, so more than one possibility exists in projective geometry. The first definitions chosen here are credited to Jacob Steiner and are used because they emphasize clearly the relationship between conics and the projectivities already studied. In the definitions, *pencils of lines* are sets of concurrent lines, and *ranges of points* are sets of collinear points.

DEFINITION. A *point conic* is the set of points that are intersections of corresponding lines in two projectively related pencils of lines in the same plane. The centers of the projective pencils must be distinct, and the sets cannot be perspective.

DEFINITION. A *line conic* is the set of lines that join corresponding points in two projectively related ranges of points in the same plane. The two lines must be distinct, and the set cannot be perspective.

It would be well to reflect on these remarkable definitions. Observe that several previous figures in this chapter have included points on a conic, although the text has not called attention to the fact. While a complete reconciliation of these definitions with the intuitive ideas of what a conic is might be extremely difficult, it is not hard to see that the definitions do provide for the fact that no three points of a point conic are collinear. Figures 7.36a and b show a point conic

and a line conic, with the connections between them and the projectivities that determine them emphasized. In Figure 7.36a, A and B are
the centers of two projectively related pencils, with C, D, E, F points of
intersection of corresponding lines. In Figure 7.36b, a and b are lines
with two projectively related ranges of points. A, A'; B, B'; C, C' are pairs
of corresponding points.

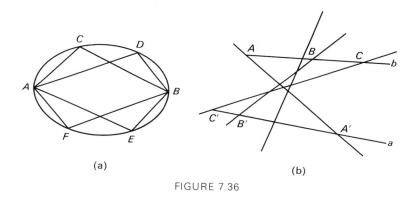

(a)   (b)

FIGURE 7.36

Two theorems stated and proved for point conics indicate
additional properties.

THEOREM 7.21. The centers of the pencils of lines in the
projectivity defining the point conic are also points of the conic.

In Figure 7.37, let A and B be the given centers. If $\overleftrightarrow{AB}$ is
considered as one of the lines in the pencil with A as center, then its
corresponding line m is one of the lines in the pencil with B as

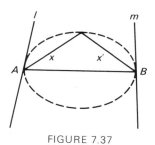

FIGURE 7.37

center. The two lines intersect at B, a point on the conic. Similarly, A lies on $\overleftrightarrow{AB}$, considered as a line through B, and on its corresponding line l through A.

THEOREM 7.22. The lines corresponding to the common line of the two pencils of lines determining a point conic are the tangents at the centers of the two pencils of lines.

DEFINITION. A *tangent* to a point conic is a line in the plane of the conic having exactly one point in common with the conic.

In Figure 7.37, each line x through A has two points in common with the conic, A and the intersection of x and its corresponding line. But by Theorem 7.21, if x' is $\overleftrightarrow{AB}$ considered as a line through B, then x and x' intersect at A, so x must be the tangent l.

A second approach to defining conics in projective geometry requires the introduction of several new concepts. Up to this point, projectivities have been defined in such a way that they pair points with points and lines with lines. It is possible to generalize the definition to include correspondence between points and lines. The original type of projectivity is a *collineation,* and the newer type now introduced is a *correlation.* The set of equations representing a correlation would have line coordinates for the primed letters but otherwise would be the same form as for a collineation. In a correlation, an element and its plane dual can correspond. Thus, the image of a complete quadrangle is a complete quadrilateral.

The projective correlation of particular interest here is a correlation of period two, called a *polarity.* A polarity pairs a point called the *pole* and a line called its *polar,* such as A and a in Figure 7.38. Any point X on a is paired with a line x on A. Points such as A and

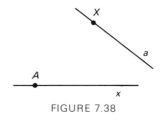

FIGURE 7.38

X are *conjugate points,* since the polar of one passes through the other. Similarly, x and a are conjugate lines. Note that the setting for the words *pole* and *polar* here is somewhat different from that used before.

In Figure 7.38, conjugate points A and X are distinct. A polarity is called a *hyperbolic polarity* if a point can be self-conjugate.

FIGURE 7.39

A self-conjugate point lies on its polar. Figure 7.39 shows various lines in a hyperbolic polarity, with the self-conjugate points indicated. These concepts can now be used to give an alternate definition for a conic.

DEFINITION. A point conic is the set of self-corresponding points in a hyperbolic polarity.

The section concludes with two major theorems, plane duals of each other, giving an additional property of any conic.

THEOREM 7.23. *Pascal's theorem.* If a simple hexagon is inscribed in a point conic, the intersection of the three pairs of opposite sides are collinear.

DEFINITION. A *simple hexagon* is a set of six points in a plane, no three collinear, and the lines joining them in a particular order.

Note that this definition allows hexagons in which the opposite sides intersect within the conic.

Let $ACBEFD$ in Figure 7.40 represent any inscribed simple hexagon. The pairs of opposite sides are $\overleftrightarrow{AC}$ and $\overleftrightarrow{EF}$, $\overleftrightarrow{CB}$ and $\overleftrightarrow{FD}$, and $\overleftrightarrow{BE}$ and $\overleftrightarrow{DA}$. Let the points of intersection be P_1, P_2, P_3. It is required to show that these three points are collinear.

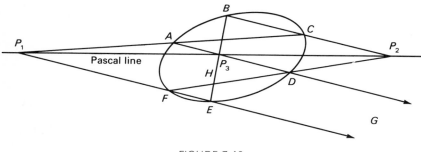

FIGURE 7.40

Two points A and B can be chosen as centers of pencils of lines in a projectivity determining the conic. There is also a projectivity established on $\overleftrightarrow{EF}$ and $\overrightarrow{FD}$. This means that if $\overrightarrow{AD}$ and $\overleftrightarrow{EF}$ meet at G, and $\overleftrightarrow{BE}$ and $\overrightarrow{FD}$ meet at H, then $P_1\, EFG \barwedge P_2\, HFD$. These two sets of points have their common element F self-corresponding, and they are related by a perspectivity (see Exercise 8, Exercise Set 7.9). The center of this perspectivity is P_3, which implies that P_1 and P_2 lie on a line through P_3, as was to be proved.

The six vertices of a simple hexagon determine sixty different hexagons, found by connecting the points in different orders. Each of these hexagons in turn has a different Pascal line. The sixty Pascal lines associated with six given points on a conic are known as *Pascal's mystic hexagram.*

Pascal's theorem was proved by Blaise Pascal (1623–1662), but the plane dual was not proved until much later, by C. J. Brianchon (1785–1864), after the development of the concept of duality.

THEOREM 7.24. *Brianchon's theorem.* If the six lines of a simple hexagon are lines of a line conic, then the three lines connecting pairs of opposite vertices of the hexagon are concurrent.

Figure 7.41 illustrates the theorem. The point of concurrency is called the Brianchon point.

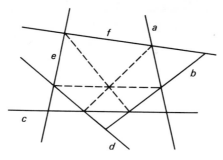

FIGURE 7.41

EXERCISES   7.9

1. What would be the effect of dropping the restriction that the pencils cannot be perspective in the definition of a point conic?

2. Point out figures in this chapter, prior to Figure 7.35, in which points on a conic could be located.

3. State the plane dual of Theorem 7.21.

4. State the plane dual of Theorem 7.22.

5. Write the equation of the line that is the image of the point (1, 1, 1) under the correlation with equations

$$X_1' = x_1 + x_2 + x_3,$$
$$X_2' = x_1 - x_2 + x_3,$$
$$X_3' = x_1 + x_2 - x_3.$$

6. Prove that the line joining two self-conjugate points in a polarity cannot be a self-conjugate line (passing through its pole).

7. In Figure 7.38, let the intersection of a and x be point Y. A triangle such that each vertex is the pole of the opposite side is called a self-polar triangle. Is triangle AXY necessarily a self-polar triangle?

8. Prove that if two sets of points in a projectivity on two lines have their common point self-corresponding, they are related by a perspectivity.

9. Explain why six points, no three collinear, determine sixty hexagons.

10. Prove Brianchon's theorem directly.

7.10   CONSTRUCTION OF CONICS

The theory of Section 7.9 has provided the basis for constructions of conics in projective geometry, using only a straightedge. The fact that conics have been defined in terms of projectivities means that any number of points or lines on a conic can be constructed using only a straight-edge, if enough information is given to determine a projectivity. This procedure is analyzed through examples.

EXAMPLE. Given five points on a point conic, construct other points.

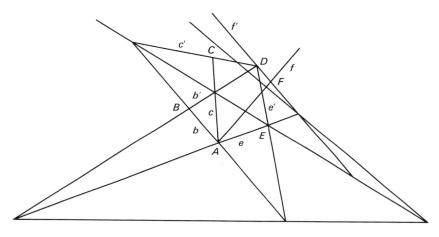

FIGURE 7.42

Let A, B, C, D, E in Figure 7.42 be the five given points. Choose any two of these, say A and D, as centers of the pencils of lines to determine the projectivity. Pairs of corresponding lines meet at B, C, E.

The lines connecting the intersections of the cross joins determine S, the center of homology.

Let f be another line through A. Since fe' and ef' must meet at S, f' can be constructed. The intersection of f and f' is the required point F.

EXAMPLE. Given four points on a conic and the tangent at one of them, construct another point of the point conic.

If points A, B, C, D, and tangent a are given as in Figure 7.43, the projectivity can be determined by the three pairs of lines shown.

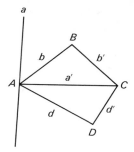

FIGURE 7.43

EXAMPLE. Construct additional lines of a line conic, given five lines, no three concurrent.

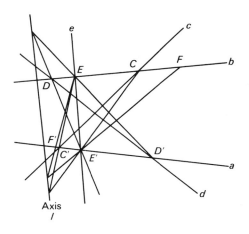

FIGURE 7.44

Let the five given lines be a, b, c, d, e in Figure 7.44. Two of these, say a and b, can be chosen as the lines to contain the corresponding points in the projectivity, while the other three determine three pairs of corresponding points. The intersections of pairs of cross joins determine the axis of homology l.

Choose any point F on b. $\overleftrightarrow{FC'}$ and $\overleftrightarrow{CF'}$ must meet on the axis so that F' can be determined. Then $\overleftrightarrow{FF'}$ is another line of the conic.

The theorems of Pascal and Brianchon, which include degenerate forms in which all six given points are not unique, provide an alternative way of constructing additional points or lines on a conic.

EXAMPLE. Find one additional point on the same conic as the five points given in Figure 7.45.

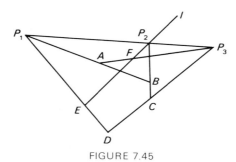

FIGURE 7.45

Let A, B, C, D, E be the five given points, with l any line through E not intersecting one of the given points. The sixth point F will be located on l so that A, B, C, D, E, F will be points of an inscribed hexagon.

The given information is enough to determine two of the points on the Pascal line and hence the line itself. Let $\overleftrightarrow{AB}$ and $\overleftrightarrow{DE}$ meet at P_1, and let $\overleftrightarrow{BC}$ and l meet at P_2. Then $\overleftrightarrow{CD}$ and $\overleftrightarrow{AF}$ must meet at a point P_3 on $\overleftrightarrow{P_1 P_2}$ so that F is the intersection of $\overrightarrow{AP_3}$ and l.

This section concludes the presentation of projective geometry as a more general geometry that includes Euclidean geometry as a special case. In the next chapter, a much more general geometry is introduced, and projective geometry will be just a special subgeometry.

EXERCISES   7.10

1. Locate five new given points in Figure 7.42, and use the method described in the text to construct a sixth point on the conic.

2. Locate new given points and a given line in Figure 7.43 and use the method described in the text to construct a sixth point on the conic.

3. Locate five new given lines in Figure 7.44 and use the method described in the text to construct a sixth line on the conic.

4. Construct another line of a conic, given four lines and the point of contact on one.

5. Use Brianchon's theorem to find one additional line of a line conic if five lines are given.

6. Use Pascal's theorem with a degenerate hexagon to construct one more point on a conic if four points and the tangent at one of these are known.

7. Given five points on a point conic, use any method with a straightedge alone to construct ten additional points on the conic.

CHAPTER 8

GEOMETRIC TOPOLOGY

8.1   TOPOLOGICAL TRANSFORMATIONS

From an intuitive (but not totally correct) point of view, topology has been thought of as "rubber sheet" geometry. Visualize sets of points on a thin sheet of rubber, as in Figure 8.1. The sheet may be stretched and twisted, but not torn or placed so that two distinct points actually coincide. Intuitively, it can be seen that the image of a circle could be an ellipse, a triangle, or a polygon, for example. Straight lines are not necessarily changed into straight lines. Most of the common properties of Euclidean geometry are no longer preserved under the set of topological transformations. Before you read ahead, it would be worth-

277

FIGURE 8.1

while to try to discover some of the properties that are preserved in topology.

Much of the basic vocabulary of topology has already been introduced, especially in Chapter 3 on convexity. Historically, topology and convexity are closely connected and share many of the same fundamental concepts, such as neighborhood of a point and interior, exterior, and boundary points.

Topology is one of the modern geometries created within the past century. Outstanding names in the history of topology include A. F. Moebius (1790–1868), J. B. Listing (1808–1882), and Bernhard Riemann (1826–1866). The study of topology continues to grow and develop, with some American mathematicians in the forefront. Courses in topology are common at the graduate level, and some are being introduced at the undergraduate level. Intuitive concepts from topology are frequently used as enrichment activities in both secondary and elementary schools.

The introduction of an intuitive idea of topology has shown the need to state and discuss a formal definition of the subject. Topology is a branch of mathematics in its own right, but the definition and discussion given here will be limited to the geometric aspects of topology.

DEFINITION. Topology is the study of those properties of a set of points invariant under the group of bicontinuous transformations of a space onto itself.

DEFINITION. A transformation f is *bicontinuous* if and only if f and f^{-1} are both continuous.

The concept of a *continuous* transformation, used extensively in calculus, is reviewed briefly, since its definition makes use of the topological concept of neighborhood of a point. A transformation of S_1

onto S_2 is continuous if for every point of S_2 and each positive number ε there is a positive number δ such that the image of any point of S_1 that is in the neighborhood of point A with a radius of δ is in the neighborhood of the image of A with a radius of ε. This idea may be stated briefly in the following notation:

$$f[N(A,\delta) \cap S_1] \subset N[f(A),\varepsilon].$$

This relationship is further clarified by a study of Figure 8.2.

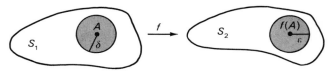

FIGURE 8.2

If the preceding paragraph seems particularly difficult, you may profit from the detailed study or review of the concept of continuity found in a modern calculus text. Intuitively, a continuous transformation takes one set of points sufficiently near each other into another set of points located near each other. Physically, the tearing apart of a surface could take close points into points at great distances.

A translation in Euclidean geometry is a very simple example of a continuous transformation. The values of ε and δ are equal, since the points in any circular region have images in another congruent circular region. Another example of a continuous transformation is $x \rightarrow x^3 + 2x - 1$, whereas an example of a transformation that is not continuous is $x \rightarrow \tan x$.

Topology may be considered as a generalization of both Euclidean geometry and projective geometry, since the group of plane motions and the group of projective transformations are both proper subgroups of the group of topological transformations. A topological transformation is sometimes called a *homeomorphism*.

THEOREM 8.1. The set of topological transformations of a space onto itself is a group of transformations.

Questions about the proof of Theorem 8.1 can be found in Exercise Set 8.1.

It was suggested earlier in the section that you think about which properties are preserved under the group of topological transformations. This section concludes with the exploration of one of these properties that is fundamental.

A basic concept in topology is that of a *connected set* of points.

DEFINITION. A set is connected if and only if any two points of the set can be joined by some curve lying wholly in the set.

All convex sets are connected, since any two of their points can be joined by segments in the set. The requirement for a set to be connected is a much looser requirement than for a set of points to be convex in the following sense: if all the curves in the definition of a connected set must be segments, then the set is convex. Figure 8.3a shows examples of connected sets that are not convex, whereas Figure 8.3b shows a set that is not connected.

(a)   (b)

FIGURE 8.3

The property of being a connected set is not itself a topological invariant. Not all connected sets of points are topologically equivalent. In other words, it is not always possible to find a topological transformation such that any two given connected sets of points are images of each other. There are different types of connectivity, however, and each specific type of connectivity is a topological invariant. This means that a set of points and its image for any topological transformation have the same type of connectivity. The analysis of types of connectivity in this section is confined to two-dimensional sets of points with interior points in a plane.

DEFINITION. If a set is *simply connected,* any closed curve in the set can be continuously deformed to a single point in the set. A connected set that is not simply connected is *multiply connected.*

An example of a simply connected set is shown in Figure 8.4a, whereas a multiply connected set is shown in Figure 8.4b.

(a)   (b)

FIGURE 8.4

What is meant by being continuously deformed to a single point can be made clearer by considering a closed curve, like a rubber band, lying entirely within the region in Figure 8.4a. The band could be shrunk without any point going outside the region. On the other hand, suppose the rubber band was in the region in Figure 8.4b, but wrapped around the hole. There would be no way to shrink it to a point without going through points in the hole. A multiply connected set, such as the one in Figure 8.4b, can be converted into a simply connected set by making one cut as shown in Figure 8.5. Verify this intuitively by considering the rubber band situation again. However, a connected set with two holes, as shown in Figure 8.6a, requires two cuts (such as those in Figure 8.6b) to convert it into a simply connected set.

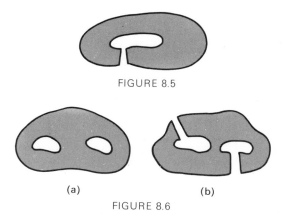

FIGURE 8.5

(a)   (b)

FIGURE 8.6

DEFINITION. In general, if $n - 1$ nonintersecting cuts are needed to convert a set into a simply connected set, the domain is *n-tuply connected.*

FIGURE 8.7

For example, the degree of connectivity is three if two cuts are necessary. The degree of connectivity of a set of points is an invariant under the group of topological transformations. For example, Figure 8.7 shows two plane sets of points that are topologically equivalent, both with degree of connectivity of four. These two sets of points are images of each other for some topological transformation. This first example of a topological invariant, degree of connectivity, is so obviously basic as to appear to have no significance in Euclidean geometry.

EXERCISES   8.1

1. Give one additional example of a continuous transformation and one additional example of a transformation that is not continuous.

Exercises 2 and 3 concern the proof of Theorem 8.1.

2. Explain why the inverse of a bicontinuous transformation is always a bicontinuous transformation.

3. Outline the proof of the fact that the product of two bicontinuous transformations is a bicontinuous transformation.

4. Name some invariants under the group of plane motions that are not topological invariants.

5. Write a definition of convex set as a special kind of connected set.

6. Which of the following sets of points are always connected sets?

  a. Line.   b. Circle.
  c. Polygon.   d. Angle.
  e. Hyperbola.   f. Reuleaux triangle.
  g. Two concentric circles.   h. Half-plane.
  i. Open set.   j. Bounded set.

7. Which of these sets are always connected sets?

  a. Sphere.   b. Hyperboloid of two sheets.
  c. Ellipsoid.   d. Tetrahedron.
  e. $\{x, y, x \mid x > y\}$.
  f. $\{x, y, z \mid x, y,$ and z are rational numbers$\}$.

8. Give a practical example of a flat object from the physical world resembling a set of points with connectivity of degree:

   a. One.   b. Two.   c. Five.

9. Draw two other sets of points topologically equivalent to those of Figure 8.7.

10. Which of these properties seem to be topological invariants?

   a. Intersection of curves.   b. Cross ratio.
   c. Midpoint.   d. Convexity.

8.2   SIMPLE CLOSED CURVES

Section 8.1 introduced topological transformations and gave an example of degree of connectivity as a topological invariant. In this geometric approach to topology, informal arguments are necessary more often than in a regular course in topology that uses analysis extensively.

The concept of simple closed curve was used in Chapter 3 on convexity. The definitions of Section 3.1 should be reviewed if necessary. The reason for extending the study of simple closed curves in this chapter is that the property of being a simple closed curve is a topological invariant. The image of a simple closed curve under a topological transformation is a simple closed curve.

In general, proving that one curve is the image of another under topological transformations is difficult. A special case is given to illustrate the method.

THEOREM 8.2. A simple closed curve that is the boundary of a two-dimensional convex body is the bicontinuous image of a circle.

In Figure 8.8, K is the convex body. K contains an interior

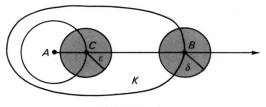

FIGURE 8.8

point A, and there is a circle with center A lying wholly within K. Each ray with endpoint A intersects the circle and the boundary of the convex body in exactly one point each, so there is a transformation of the boundary onto the circle. This transformation must be shown to be continuous. Let B and C be corresponding points, one on the boundary of K and the other on the circle. For ε any positive number, a number δ can be found small enough so that the segment joining any point of $N(B, \delta)$ to A will intersect the circle in a point of $N(C, \varepsilon)$. Then the transformation is continuous.

For the transformation to be bicontinuous, the inverse transformation must also be continuous. In Figure 8.9, ε is any positive number representing the radius of the neighborhood of point B, on the boundary of the convex body.

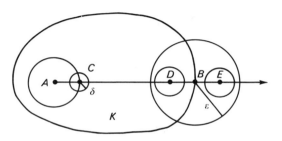

FIGURE 8.9

On ray $\overrightarrow{AB}$, it is possible to find a point D with neighborhood $N(D, \alpha)$ in the interior of both K and $N(B, \varepsilon)$ and an exterior point E with neighborhood $N(E, \beta)$ in the exterior of K but in the interior of $N(B, \varepsilon)$, for some positive numbers α and β.

Now, δ can be chosen small enough for $N(C, \delta)$ that any ray with endpoint A passing through a point of $N(C, \delta)$ will also pass through both $N(D, \alpha)$ and $N(E, \beta)$. Since one of the neighborhoods is interior to K and one is exterior, a boundary point of K lies on the ray and also lies in $N(B, \varepsilon)$. It follows that the inverse transformation meets the definition of a continuous function.

Some proofs of the fundamental theorem of algebra depend on an application of topology using ideas very similar to those in the proof of Theorem 8.2.

Consider the intuitive notion of tracing a simple closed curve by moving a pencil over a piece of paper, subject only to the restriction that you cannot cross a previous path and that you must return to the starting point. The result might be a complicated drawing such as the one in Figure 8.10. It is hard to distinguish between interior and

FIGURE 8.10

exterior points, let alone to visualize a simple interior. A significant theorem about simple closed curves proved in topology, but assumed in earlier chapters and in elementary geometry, is called the Jordan curve theorem.

THEOREM 8.3. *Jordan curve theorem.* Any simple closed curve in the plane partitions the plane into three disjoint connected sets such that the set that is the curve is the boundary of both the other sets.

The Jordan curve theorem is proved in this section only for the following special case in which the simple closed curves are polygons.

THEOREM 8.4. Any simple closed polygon in the plane partitions the plane into three disjoint connected sets such that the set that is the polygon is the boundary of the other two sets.

Let $\overrightarrow{OX}$ be some fixed ray in the plane, not parallel to one of the sides of the polygon. Figure 8.11 shows a typical example. Any point P of the plane may be regarded as the endpoint of a ray parallel to $\overrightarrow{OX}$. In Figure 8.11, three such rays are shown, with endpoints P, P', and P''. The number of points of intersection of the rays and the polygon are one, three, and two, respectively. If the number of intersections is odd, the endpoint P is said to have an *odd*

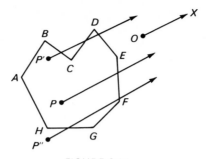

FIGURE 8.11

parity; if the number of intersections is even, the endpoint has an *even parity*. Thus, P and P' have an odd parity, while P'' has an even parity.

To find the parity of each point in the plane, it is necessary to consider the possibility of a ray passing through one or more vertices of the polygon. Two possible relationships of adjacent sides and the vertex of intersection are illustrated in Figure 8.12. In Figure 8.12a

(a)   (b)

FIGURE 8.12

the vertex is not counted as an intersection. For Figure 8.12b the vertex is counted. The vertex is not counted when the two adjacent sides of the polygon are on the same side of the ray, but it is counted when they are on opposite sides of the ray.

Let set S be the set of all points in the plane not on the polygon such that the parity is odd, and let S' be the set of all points in the plane not on the polygon such that the parity is even. Let $\overline{PP'}$ be any segment not intersecting the polygon. Then every point on $\overline{PP'}$ has the same parity, and the segment lies entirely in S or S'.

Now it can be shown that any two points P, P' of the same

FIGURE 8.13

set S or S' can be joined by a polygonal curve that does not intersect the polygon. If $\overline{PP'}$ does not intersect the polygon, the statement is obvious. Otherwise, let X be the point of intersection closest to P, and let X' be the point of intersection closest to P', as shown in Figure 8.13. Let P'' be a point of $\overline{PP'}$ near P' and with the same parity as P'.

It is possible to trace a polygonal path with sides close to the sides of the original polygon until a point P''' on $\overline{PP'}$ near P is reached. P''' is between P and X, and the polygonal path from P'' to P''' is between P and X, and the polygonal path from P'' to P''' is entirely composed of points with the same parity.

The original polygon is the boundary of both set S and set S'. If A is any point of the polygon, then there are points B and C on either side of A, but arbitrarily close, such that their parity is different. Since every neighborhood of A contains points of both S and S', the polygon is the boundary of both S and S'. Points in the exterior of the polygon have an even parity, and points in the interior have an odd parity.

The following two theorems, whose proofs are left as Exercises 6 and 7 of Exercise Set 8.2, give additional properties of simple closed curves and their intersections with one-dimensional convex sets in the same plane.

THEOREM 8.5. For one point A in the interior and one point B in the exterior of a simple closed curve S, $\overline{AB} \cap S$ is not empty.

THEOREM 8.6. Every ray with an endpoint in the interior of a simple closed curve intersects the curve.

The Jordan curve theorem is involved in various puzzle-type problems in topology. One of these is illustrated in Figure 8.14a.

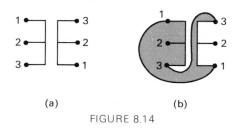

(a)   (b)

FIGURE 8.14

The problem is to draw curves connecting the three pairs of points with the same numbers in such a way that the curves do not intersect each other or any of the other lines given in the figure. The problem is impossible to solve, as shown by Figure 8.14b. The given lines and the paths drawn from 1–1 and 3–3 result in a shaded region representing a simple closed curve and its interior. One of the points labeled 2 is an interior point and the other is an exterior point, so any curve connecting them must intersect the boundary.

EXERCISES   8.2

1. Which of these sets of points are topologically equivalent to any simple closed curve?

   a. Triangle.   b. Hyperbola.
   c. Two concentric circles.   d. Boundary of Reuleaux triangle.

2. Which of these plane regions are topologically equivalent to any simple closed curve and its interior?

   a. Triangular region.   b. Circular region.
   c. Plane region with degree of connectivity one.
   d. Plane region with degree of connectivity two.

3. In Figure 8.11, explain what happens when vertices C, E of the polygon are encountered by rays drawn parallel to $\overrightarrow{OX}$ with points on $\overline{P'P}$ as endpoints.

4. Do the points in the interior of a simple closed curve have odd or even parity?

5. Is the exterior of a simple closed polygon simply connected or multiply connected?

6. Prove Theorem 8.5.

7. Prove Theorem 8.6.

8. In Figure 8.15, can curves be drawn connecting the three pairs of points with the same numbers in such a way that the curves do not intersect each other or any of the other lines given in the figure? Explain why or why not.

FIGURE 8.15

9. Pasch's axiom for Euclidean geometry states that a line in the plane of a triangle which intersects one side of the triangle at a point other than the vertex also intersects a second side. Prove Pasch's axiom as a theorem in topology, using the theorems of this section.

8.3   INVARIANT POINTS AND NETWORKS

The study of invariants under the group of topological transformations is continued in this section. In the study of motions of a plane, invariant points were common, but for more general transformations, they are more uncommon. One of the simplest examples of a theorem about invariant points under topological transformations is the Brouwer fixed point theorem for a circular region. This theorem is named for L. E. J. Brouwer, a famous twentieth-century Dutch mathematician.

THEOREM 8.7. *Brouwer fixed point theorem*. If the points of a circular region undergo a continuous transformation so that each image is a member of the set, then there is at least one fixed point.

A simple example of Theorem 8.7 is a circle rotated about its center, with the center the fixed point. The theorem can be proved indirectly. Assume that there is no fixed point. The transformation may be visualized by associating a vector with each point, as in Figure 8.16a. The initial point of each vector is the original point, and the terminal point is its image.

All of the vectors for points on the boundary of the region

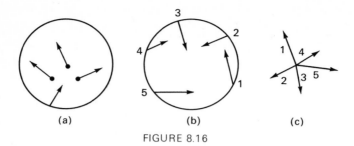

FIGURE 8.16

point into the circle, as in Figure 8.16b. Suppose that the points on the boundary are considered in a counterclockwise order around the circle. The use of a vector diagram as in Figure 8.16c shows that the vector turns around and comes back to its original position. It can be shown that the total algebraic change in the direction of the vectors (not the vector sum) amounts to one positive revolution (the *index* of the vectors is one). This can be shown by considering the tangent vectors for all points on the circle, as in Figure 8.17.

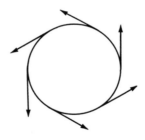

FIGURE 8.17

The tangent vectors make one complete revolution, and if the vectors for the points on the circumference turn through some different angle, the difference is a multiple of 2π. This means the vectors of the transformation must turn completely around the tangents at least once. But since the vectors turn continuously, at some time the transformation vector and the tangent vector must point in the same direction. This is impossible, since all the transformation vectors point inside the circle.

For any circle concentric to the given circle and contained within it, the index of the transformation vectors must also be one. This is true because passing continuously from one circumference to smaller ones results in a continuous change in the transformation vector index. This means that the changes cannot take place, since the vector index may assume only integral values. No matter how small the concentric circle is, the index of the vectors of the transformation will always be one. But this is impossible; the vectors on a sufficiently small circle will all point in approximately the same direction as the vector at the center of the circle, since the transformation is continuous. If the index of the vectors remains an integer and becomes smaller than one, it will be zero. Thus, the assumption of a vector for every point has led to a contradiction, so there is at least one fixed point.

A recent application of invariant points and topology is in the field of oceanography. In an attempt to explain the theory of continental drift, a mathematical model is used to illustrate what happens when two rigid plates on a sphere spread out from a ridge crossed by fracture zones. The plates must revolve about an invariant point called the *pole of spreading*. The rotation also takes place around an *axis of spreading*, which passes through the pole of spreading and the center of the earth. For more information, read "The Origin of the Oceans," by Sir Edward Bullard, in the September, 1969, *Scientific American.*

Another application of the concept of a fixed point theorem from topology is in one proof of the fundamental theorem of algebra. Recall that this theorem states that any polynomial equation with complex coefficients has a root in the field of complex numbers.

The second example of a topological invariant in this section is related to the idea of a *network*. The word "network" is used to describe a connected set of vertices, segments, and portions of curves such as those shown in Figure 8.18. These may be thought of as a set

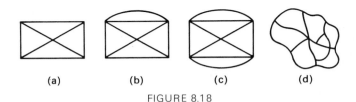

(a)   (b)   (c)   (d)

FIGURE 8.18

of vertices with paths connecting them. Networks such as these furnish examples of what are known as *tracing puzzles*. Can the networks be traced without lifting the pencil from the paper and without repeating a portion (that is, not more than a finite set of isolated points) of the path? Try to trace the networks in Figure 8.18a–c before reading the next paragraph.

The network in Figure 8.18a cannot be traced, whereas those in Figures 8.18b and c can be traced. Try to decide what is the essential difference between networks that can be traced and those that cannot before you read the next paragraph.

The number of paths leading to each vertex in a network is crucial in trying to decide whether or not the network can be traced. If a vertex has an even number of paths to it (an *even vertex*), then these may be used in pairs in going to and leaving the vertex. If there are an odd number of paths to a point (an odd vertex), they cannot be used in pairs, and it is necessary to begin or end at that point. If the network has two or fewer vertices with an odd number of paths, it can be traced. The network will be a curve in this case. If there are no vertices with an odd number of paths, then the network can be traced by beginning and ending at the same point. The network will be a closed curve in this case. Now check Figure 8.18d. This network has more than two vertices with an odd number of paths, so it cannot be traced. The information about tracing puzzles is summarized in the following table.

TABLE 8.1

Number of Vertices with Odd Number of Paths	Network Can Be Traced	Name of Network
more than two	no	not a curve
two or fewer	yes	curve
zero	yes	closed curve

The essential characteristics of a network in terms of the number of even and odd vertices is a topological invariant. In other words, for

FIGURE 8.19

a particular network, the number of paths leading to a vertex is not changed by a topological transformation. For example, the two networks in Figure 8.19 are topologically equivalent.

EXERCISES   8.3

1. Could the continuous transformation in Theorem 8.7 be a glide reflection? Why?

2. Outline briefly the major steps in the proof of Theorem 8.7.

For Exercises 3–7, tell whether or not the networks in Figure 8.20 can be traced.

3.   4.   5.   6.   7.

FIGURE 8.20

8.   Which of the networks in Exercises 3–7 are curves?

9.   Which of the networks in Exercises 3–7 are closed curves?

10.   Which of the networks in Exercises 3–7 are simple closed curves?

8.4   INTRODUCTION TO THE TOPOLOGY OF SURFACES

The topology of Sections 8.1–8.3 was the topology of the plane. The last two sections include concepts and theorems from the topology of surfaces in three dimensions. This section introduces some basic theorems for simple closed surfaces, analyzes a topological invariant of surfaces, and gives an example of an unsolved problem in topology.

The following theorem is the Jordan curve theorem stated for three-space.

THEOREM 8.8. Any simple closed surface in three-space partitions the space into three disjoint connected sets such that the set that is the surface is the boundary of both of the other sets.

The following theorem and corollary can also be proved for simple closed surfaces; they are analogous to the theorem proved for the simple closed curve.

THEOREM 8.9. For one point A in the interior and one point B in the exterior of a simple closed surface S, $\overline{AB} \cap S$ is not empty (see Figure 8.21).

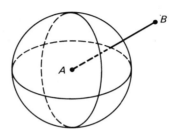

FIGURE 8.21

The following corollary is true because the simple closed surface is bounded; hence the ray must have an exterior point.

THEOREM 8.10. Every ray with an endpoint in the interior of a simple closed surface intersects the surface.

The property of being a simple closed surface is an invariant under the group of topological transformations in three-space.

Corresponding to the study of connectivity for plane regions is the analysis of the *genus* of a surface.

DEFINITION. The genus of a surface is the largest number of nonintersecting simple closed curves that can be drawn on the surface without separating it into two unconnected parts.

For example, the genus of a sphere is zero, since any simple closed curve separates it into two unconnected parts, as shown in

Figure 8.22. Similarly, many common surfaces such as an ellipsoid and a convex polyhedron are also of genus zero.

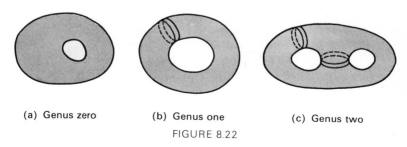

(a) Genus zero   (b) Genus one   (c) Genus two

FIGURE 8.22

Other surfaces exist besides those with a genus of zero. Figure 8.22b shows a *torus* with genus one, and Figure 8.22c shows a surface with genus two. For the figure of genus one, a second cut would separate the surface into two unconnected parts. Similarly, for the figure of genus two, a third cut would separate the surface into unconnected parts.

The genus of a surface is another example of an invariant under any topological transformation. Also, any surface of a particular genus may be changed into any other surface of the same genus by a topological transformation. The effect of the two previous statements is that two surfaces with the same genus are topologically equivalent. Figure 8.23 shows several topologically equivalent surfaces of genus one.

FIGURE 8.23

An interesting and unsolved topological problem closely associated with networks, with simple closed curves, and with invariance concerns the coloring of maps. The *map-coloring problem* has been such a common enrichment topic in mathematics that many students may be familiar with it to some extent; nevertheless, it is a current topic of research in topology.

FIGURE 8.24

For two-dimensional maps such as those shown in Figure 8.24 or for similar ones on a sphere, only four colors are necessary if each two countries with a common boundary must have a different color. It is assumed that two countries may have the same color if the boundaries intersect in a single point. This is the case for the two countries numbered 1 in Figure 8.25.

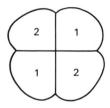

FIGURE 8.25

Strangely enough, the general statement that any map can be colored with four colors has not been proved, although Moebius proposed proving the statement as early as 1840. About 50 years later, the following theorem was proved.

THEOREM 8.11. Every map on a sphere can be colored according to the rules for the map-coloring problem by using at most five colors.

The countries in Theorem 8.11 are assumed to be simply connected regions. The rules for the map-coloring problem are that two countries with a common boundary must have a different color but that they may have the same color if the boundaries intersect in a single point.

As was pointed out at the beginning of the discussion, the map-coloring problem is related to other topological concepts. The boundaries of the countries can be any simple closed curve, so the exact shape is unimportant. The map-coloring problem changes if the genus of the surface changes. Theorem 8.11 applies to any surface of genus zero, but not to a torus, for example. Special cases of the four-color map problem, for up to some certain number of countries, have been proved, but the general problem continues to interest many topologists.

EXERCISES   8.4

1. Prove Theorem 8.9.
2. What is the genus of a simple closed surface?
3. Name or describe other surfaces of genus zero, one, and two besides those mentioned in this section.
4. Show that the map in Figure 8.26 can be colored with not more than four colors, according to the restrictions of the map-coloring problem.

FIGURE 8.26

ıt is the minimum number of colors required to color each map in ıre 8.24 according to the rules of the map-coloring problem?

w a map with 12 countries that requires only three colors, yet stays in the map-coloring requirements.

w a map with 12 countries that requires only two colors to fulfill the ı-coloring requirements.

lain why it is important that the countries on a map in the map- ring problem be simply connected regions.

lain why it is important in the map-coloring problem that countries with a le point in common can have the same color.

8.5   EULER'S FORMULA AND SPECIAL SURFACES

This section continues the study of the topology of surfaces by introducing Euler's formula for surfaces of different genus and by suggesting special surfaces with strange properties unlike those of ordinary Euclidean geometry.

The development of the subject of topology has come during the past one hundred years, but it did have its beginning before that. One of the earliest ideas that is actually topological in nature was first discovered by Descartes, then rediscovered by Euler in 1752. It has since gone by the name of *Euler's formula.* Euler's formula relates the number of faces, vertices, and edges of a simple polyhedron. Recall that a *polyhedron* is a closed surface consisting of a number of faces, each of which is a polygonal region. If the surface has no holes in it and can be deformed into a sphere by a continuous transformation, it is a *simple* polyhedron. A simple polyhedron has a genus of zero.

The Greeks showed particular interest in five polyhedra, those that were *regular,* with congruent faces and angles. The five regular polyhedra are shown in Figure 8.27. The names and some of the

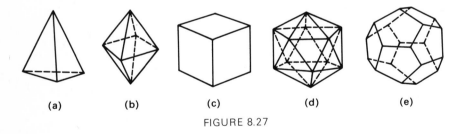

(a)   (b)   (c)   (d)   (e)

FIGURE 8.27

numbers of vertices, faces, and edges are listed in Table 8.2. The completion of the table is left as Exercise 1, Exercise Set 8.5.

TABLE 8.2

Figure Number	Name of Polyhedron	V	F	E
8.27a	tetrahedron	4	4	6
8.27b	octahedron	6	8	12
8.27c	cube	8	?	?
·8.27d	dodecahedron	?	12	?
8.27e	icosahedron	?	?	?

What is the relationship between the number of vertices, faces, and edges for these polyhedra? Take various other examples of polyhedra, if necessary, to arrive at Euler's formula on your own before continuing.

THEOREM 8.12. For a simple closed polyhedron, $V + F = E + 2$, where V is the number of vertices, E the number of edges, and F the number of faces.

To prove the formula, think of cutting out one of the faces of the polyhedron so the remaining surface can be stretched out flat by a topological transformation. The network of vertices and edges in the plane will have the same number of vertices and edges as in the original polyhedron, but there is one less face, since one has been removed. Figure 8.28 shows an example, using a cube.

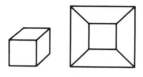

FIGURE 8.28

Now triangulate the plane network by drawing diagonals connecting vertices until a triangular decomposition has been achieved. In this triangulation, the value of $V - E + F$ is not changed, since drawing each diagonal adds one edge and one face.

Once the triangulation has been completed, the triangles may be removed one at a time until a single one remains. (This process is illustrated for the cube in Figure 8.29.) Some of the triangles will have

FIGURE 8.29

edges on the boundary of the network. First, remove any edge of a boundary triangle not an edge of another triangle.

1. If the triangle has only one edge on the boundary, removing that one edge reduces both E and F by one, so $V - E + F$ is not changed.
2. If the triangle has two edges on the boundary, removing them reduces V by one, E by two, and F by one, so $V - E + F$ is unchanged.

By a continuation of this process, the boundary can be changed until all that is left of the triangulation network is a single triangle with three edges, three vertices, and one face. For this triangle, $V - E + F = 1$. But since $V - E + F$ is an invariant in this process of removing triangles, the formula $V - E + F = 1$ holds for the original plane network. The formula $V - E + F = 2$ applies to the original polyhedron, which had one more face.

Euler's formula $V - E + F = 2$ is an example of a topological property of a figure, since it is unchanged under a topological transformation. Of course, the polyhedron can be transformed into any simple closed surface so that the system of vertices, faces, and edges becomes a network of points, regions, and paths on the surface.

Euler's formula does not hold unless the genus of the surface is zero. Furthermore, an agreement on how to count the faces is necessary. For example, the set of points in Figure 8.30 has genus one. To count the faces requires an agreement that face $AGFD$ is counted but face $IJKL$ is then subtracted so that the total number of faces is eight. Check to see that Euler's formula does not hold for this surface.

FIGURE 8.30

It is interesting to realize that the concept of the Euler formula can be generalized for any closed surface, regardless of genus.

THEOREM 8.13. If a closed surface of genus n is partitioned into regions by a number of vertices joined by curved arcs, then $V - E + F = 2 - 2n$ (F represents regions and E the arcs of the network). The number $2 - 2n$ is called the *Euler characteristic*. Note that this reduces to the ordinary Euler formula for a surface of genus zero.

The final topological topic to be introduced is that of a very special kind of surface unlike those studied up to this time. Ordinary surfaces have two sides, but Moebius was the first to discover that it is possible to have surfaces with only one side. For a surface of one side, it is possible to move from any point on the surface to any other without going over the edge.

FIGURE 8.31

In Figure 8.31, which shows a surface of two sides, you could not draw a line from A to B, for example, without crossing over an edge.

The simplest one-sided surface is called a *Moebius strip*. It is pictured in Figure 8.32. A representation of a Moebius strip can be

FIGURE 8.32

formed from a rectangular strip of paper. Give the paper a half-twist and tape the ends together. A Moebius strip has only one side and one edge. You can trace a curve from any one point on the surface to any other points without crossing an edge. It is fascinating to experiment with a Moebius strip to see that it behaves unlike a two-sided surface. For example, if a Moebius strip is cut down a line through the middle, it remains in one piece.

A second example of a one-sided surface is a Klein bottle, shown in Figure 8.33. This intersecting surface has no inside or outside.

FIGURE 8.33   FIGURE 8.34

An unusual application of one-sided surfaces for women's dresses appeared in Jean J. Pedersen's article, "Dressing up Mathematics," in *The Mathematics Teacher,* February, 1968. A possibly more practical application of the Moebius strip is shown in Figure 8.34. The shaping of a belt in the form of a Moebius strip connecting two wheels allows the belt to wear out at the same rate everywhere, not just on one side as in the usual arrangement. However, it also reverses the direction of revolution.

The property of being a two-sided or a one-sided surface is a topological invariant. For example, a Moebius strip cannot be changed to a two-sided surface by a topological transformation. A related application of topology that should be mentioned briefly here is the classification of various kinds of *knots.* Two knots are topologically equivalent if one can be deformed into the other in a continuous way.

EXERCISES   8.5

1. Complete Table 8.2 showing the number of vertices, faces, and edges for the five regular polyhedra.

2. Verify Euler's formula for each of the five regular polyhedra.

3. Verify Euler's formula for a:

    a. Hexagonal right prism.   b. Octagonal right prism.

4. Make drawings similar to Figures 8.28 and 8.29 for a tetrahedron.

5. Find the Euler characteristic for the set of points in Figure 8.30.

6. What is the numerical value of the Euler constant for a surface of:

    a. Genus zero?   b. Genus one?
    c. Genus two?   d. Genus three?

7. Draw a figure similar to Figure 8.30, but with a genus of two. Find the number of vertices, faces, and edges, then determine the Euler characteristic.

8. By constructing a model, verify that if a Moebius strip is cut down a line through the middle, it remains in one piece.

9. What is the result when a Moebius strip is cut lengthwise, beginning one third of the way from the edge, rather than down the middle?

10. Suppose a surface is constructed by giving a strip of paper a full twist before gluing the ends. Is this a Moebius strip? What happens when the band is cut down the middle lengthwise?

11. Answer the same questions as in Exercise 10, but assume the strip of paper is given one and a half twists before gluing the ends.

12. A Moebius strip is a set of points in n-space, for n equal to what number?

13. A Klein bottle is a set of points in n-space, for n equal to what number?

CHAPTER 9

NON-EUCLIDEAN GEOMETRY

9.1 INTRODUCTION TO HYPERBOLIC GEOMETRY

The term *non-Euclidean geometry* is used in a very restricted sense. It does not designate any geometry that is not identical to the geometry of Euclid. Non-Euclidean geometry differs from the geometry of Euclid because it substitutes another alternative for his so-called fifth postulate. Thus, projective geometry and topology are not examples of non-Euclidean geometry in this technical sense.

Recall from Chapter 1 that the form of the fifth postulate as stated by Euclid was considerably more complex than the form of the other postulates and axioms: "If a transversal falls on two lines in such a way that the interior angles on one side of the transversal are less than two right angles, then the lines meet on that side on which the angles are less than two right angles." According to Euclid, $\overleftrightarrow{AD}$ and $\overleftrightarrow{BC}$ meet to the right in Figure 9.1 if the sum of the measures of $\angle DAB$ and $\angle ABC$ is less than π radians.

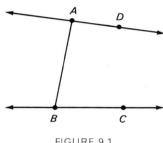

FIGURE 9.1

Note that the word "parallel" does not appear in the fifth postulate.

The wording of Euclid's fifth postulate that is most commonly used is called *Playfair's axiom*: Through a point not on a given line, exactly one line can be drawn in the plane parallel to the given line. The word "parallel" as used here means not intersecting or having no Euclidean point in common. Playfair's axiom and the original fifth postulate are logically equivalent. This means that either one can be used, along with the other assumptions of Euclidean geometry, to prove the second.

Non-Euclidean geometry provided Einstein with a suitable model for his work on relativity. While it also has applications in differential geometry and elsewhere, it is worthwhile for other reasons as well. The real understanding of the concept of postulate sometimes comes only when a person begins with postulates that are not self-evident. In non-Euclidean geometry, it is generally not possible to rely on intuition or on drawings to the same extent as is true for Euclidean geometry. Finally, non-Euclidean geometry, unlike projective geometry or topology, is significant in that it is something other than a generalization of Euclidean geometry. Non-Euclidean geometry does not include ordinary geometry as a special case.

From the time Euclid stated his postulates, about 300 B.C., mathematicians attempted to show that the fifth postulate was actually a theorem that could be proved from the other postulates. None of these people succeeded. Shortly after 1800, mathematicians such as Carl Friedrich Gauss began to realize that the fifth postulate could never be proved from the others, because it was indeed an independent postulate in the set of Euclidean postulates, not a theorem. Attempts to prove the fifth postulate by denying it had already produced strange theorems that had to be accepted as valid if some other substitute postulate was actually possible.

As in the discovery of calculus, more than one person shares the credit for the actual discovery of non-Euclidean geometry. Though Gauss was aware of the significance of the subject, he did not publish any material. The first account of non-Euclidean geometry to be published was based on the assumption that, through a point not on a given line, more than one line can be drawn parallel to a given line in the plane. This type of geometry, called *hyperbolic* non-Euclidean geometry, was discovered independently by a Russian, Nikolai Lobachewsky (1793–1856), and, at about the same time, by a Hungarian, Johann Bolyai (1802–1860). The results were published about 1830. A second type of non-Euclidean geometry, *elliptic geometry,* is introduced briefly in Section 9.6.

The development of hyperbolic geometry in this chapter is based on all the assumptions and undefined terms of modern Euclidean geometry except for the following substitution for the parallel postulate, identified as the characteristic postulate of hyperbolic geometry.

CHARACTERISTIC POSTULATE. Through a given point C, not on a given line $\overleftrightarrow{AB}$, passes more than one line in the plane not intersecting the given line.

The relationship described in the characteristic postulate is pictured in Figure 9.2, if it is assumed that $\overrightarrow{CD}$ and $\overleftrightarrow{CE}$ are two distinct lines through C and that neither intersects $\overleftrightarrow{AB}$.

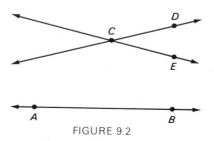

FIGURE 9.2

The idea of changing the parallel postulate may seem extremely strange, but there are many finite geometries where the parallel postulate of Euclid does not hold. For example, several of the finite geometries of Chapter 1 have an axiom stating that each two lines intersect at a point. The finite geometry of Desargues has a peculiarity concerning parallels; in that geometry, each point can have three lines through it parallel to one particular line, the polar of the point.

There is no finite geometry that has all the axioms of hyperbolic geometry, but some do have the characteristic postulate. A particular example is the geometry of thirteen points and twenty-six lines, represented by the table below, in which each set of three points lies on a line.

TABLE 9.1

A, B, C	B, D, F	C, D, G	D, H, I	E, F, K	F, G, M	G, I, K	H, J, L
A, D, E	B, E, I	C, E, J	D, J, K	E, G, L	F, I, J		
A, F, H	B, G, H	C, F, L	D, L, M	E, H, M			
A, G, J	B, J, M	C, H, K					
A, I, L	B, K, L	C, I, M					
A, K, M							

Returning to hyperbolic geometry and its characteristic postulate, you should pause to consider some of the consequences of this change in postulates before reading the development of theorems and proofs for the new plane geometry.

THEOREM 9.1. Through a given point C, not on a given line $\overleftrightarrow{AB}$, pass an infinite number of lines not intersecting the given line.

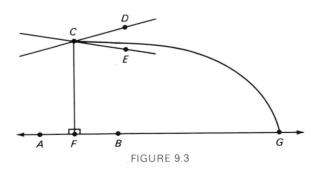

FIGURE 9.3

There are an infinite number of lines passing through C and the interior of angle DCE in Figure 9.2. For Figure 9.3, let $\overline{CF}$ be the perpendicular from C, and assume that $\overleftrightarrow{CG}$, any one of these interior lines, does intersect $\overleftrightarrow{AB}$. This means that $\overleftrightarrow{CE}$ must also intersect $\overleftrightarrow{FG}$, by the axiom of Pasch, which continues to hold in hyperbolic geometry since only the axiom of parallels from Euclidean geometry has been replaced.

Recall that the axiom of Pasch states that a line entering a triangle at a vertex intersects the opposite side. In Figure 9.3, $\overrightarrow{CE}$ enters triangle CFG. The fact that $\overleftrightarrow{CE}$ intersects $\overleftrightarrow{AB}$ contradicts the assumption that it is parallel to $\overleftrightarrow{AB}$, so $\overleftrightarrow{CG}$ cannot intersect $\overleftrightarrow{AB}$.

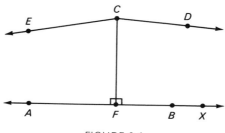

FIGURE 9.4

In Figure 9.4, the set of all lines in a plane passing through C is partitioned into two subsets, those that do intersect $\overleftrightarrow{AB}$ and those that do not. Because of the assumption of the one-to-one correspondence between sets of real numbers and sets of lines through a point, which was retained from Euclidean geometry, it is known that this partitioning is brought about by two different lines, shown as $\overleftrightarrow{CD}$ and $\overleftrightarrow{CE}$. These two lines must be either the last lines in either direction that do intersect $\overleftrightarrow{AB}$ or the first lines in either direction that do not intersect $\overleftrightarrow{AB}$. The assumption that there is a last intersecting line, say $\overleftrightarrow{CX}$, for example, is immediately contradicted by the fact that other points on $\overleftrightarrow{AB}$ to the right of X also determine intersecting lines. Then $\overleftrightarrow{CD}$ and $\overleftrightarrow{CE}$ are the first lines that do not intersect $\overleftrightarrow{AB}$.

DEFINITION. In hyperbolic geometry, the first lines in either direction through a point that do not intersect a given line are *parallel lines*.

DEFINITION. All other lines through a point not intersecting the given line, other than the two parallel lines, are *nonintersecting lines.*

According to these technical definitions, there are exactly two parallels through C to $\overleftrightarrow{AB}$. $\overleftrightarrow{CD}$ is called the *right hand parallel* and $\overrightarrow{CE}$ is the *left hand parallel.* The angles FCD and FCE are *angles of parallelism* for the distance FC.

THEOREM 9.2. The two angles of parallelism for the same distance are congruent and acute.

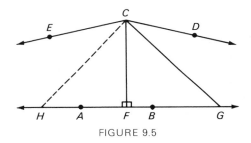

FIGURE 9.5

Assume that $\angle FCE$ and $\angle FCD$ in Figure 9.5 are angles of parallelism for CF, but are not congruent. Assume next that $\angle FCD$, for example, is greater. Then there is an angle, $\angle FCG$, congruent to $\angle FCE$ and such that $\overleftrightarrow{CG}$ is in the interior of $\angle FCD$. If $FH = FG$, then $\triangle FCG \cong \triangle FCH$, so $\angle FCH \cong \angle FCE$. This is a contradiction because $\overleftrightarrow{CE}$ has no point in common with $\overleftrightarrow{AB}$. Hence, $\angle FCD \cong \angle FCE$.

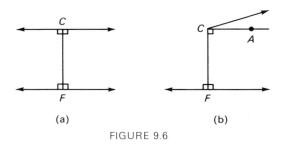

(a)   (b)

FIGURE 9.6

The angles of parallelism cannot be right angles, as in Figure 9.6a, because the two lines shown are nonintersecting. Euclid proved that fact prior to using the fifth postulate in Book 1 of his *Elements*. Furthermore, the angle of parallelism cannot be obtuse, as in Figure 9.6b, because then it would have a nonintersecting line $\overleftrightarrow{CA}$ within the angle. This would contradict the fact that the parallel is defined to be the first noncutting line.

The angles of parallelism in hyperbolic geometry are neither right nor obtuse, so they must be acute. This illustrates a consequence of the characteristic postulate that is radically different from the statement for Euclidean angles of parallelism whose measures have a sum of π.

EXERCISES   9.1

1. Prove Playfair's axiom, assuming Euclid's fifth postulate in the original form.

2. Prove the original statement of Euclid's fifth postulate, assuming Playfair's axiom.

It can be proved that each of the statements in Exercises 3–8 is equivalent to Euclid's fifth postulate. Reword each sentence so that it becomes a valid statement in non-Euclidean geometry.

3. If a straight line intersects one of two parallel lines, it will always intersect the other.

4. Straight lines parallel to the same straight line are always parallel to one another.

5. There exists one triangle for which the sum of the measures of the angles is π radians.

6. There exists a pair of similar but noncongruent triangles.

7. There exists a pair of straight lines the same distance apart at every point.

8. It is always possible to pass a circle through three noncollinear points.

9. For which of these finite geometries of Chapter 1 does Euclid's fifth postulate always hold?

    a. Geometry of Pappus.   b. Fano's geometry.
    c. Four-line geometry.   d. Geometry of Desargues.

10. For the thirteen-point finite geometry of this section, name all the lines through point A that do not have a point in common with line BDF.

11. Without using an axiom of parallelism, prove that if a transversal of two lines makes the alternate angles congruent, then the two given lines do not intersect.

12. In hyperbolic geometry, through a given point not on a given line, exactly how many lines can be drawn in that plane that are parallel to the given line?

9.2   IDEAL POINTS AND OMEGA TRIANGLES

In hyperbolic geometry, as in the geometry of the extended Euclidean plane and in projective or inversive geometry, the inclusion of ideal points is an important idea. In hyperbolic geometry, two parallel lines do not have an ordinary point in common, but they are said to meet at an ideal point.

DEFINITION. An *ideal point* in hyperbolic geometry is the point of intersection of two parallel lines.

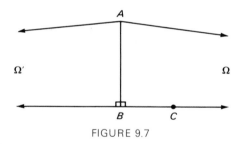

FIGURE 9.7

In Figure 9.7, the right and left hand parallels shown through point A to line $\overleftrightarrow{BC}$ meet that line in ideal points Ω (omega) and Ω'. It can be proved (see Exercise 1, Exercise Set 9.2) that Ω and Ω' are distinct points. An ordinary line in affine geometry has exactly one ideal point, but a line in hyperbolic geometry has two distinct ideal points.

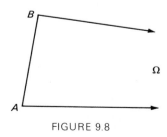

FIGURE 9.8

The development of the properties of parallel lines in hyperbolic geometry is continued by considering the *omega triangle,* a three-sided figure as in Figure 9.8, with one ideal vertex. Though not a triangle in the ordinary sense, an omega triangle does have some of the same properties as a triangle with three ordinary vertices.

THEOREM 9.3. The axiom of Pasch holds for an omega triangle, whether the line enters at a vertex or at a point not a vertex.

In Figure 9.9, let C be any interior point of the omega triangle $AB\Omega$. Then $\overleftrightarrow{BC}$ and $\overleftrightarrow{AC}$ intersect the opposite side because $\overleftrightarrow{B\Omega}$ is the first noncutting line through B for $\overleftrightarrow{AD}$ and $\overleftrightarrow{AD}$ is the first noncutting line through A for $\overleftrightarrow{BE}$. If a line $\overleftrightarrow{C\Omega}$ enters the omega triangle through the ideal vertex, it intersects $\overleftrightarrow{AB}$, because of Pasch's axiom applied to $\triangle ABD$. The second part of the proof is left as Exercise 5 of Exercise Set 9.2.

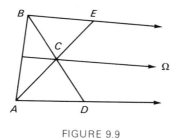

FIGURE 9.9

Euclid proved, without the use of the fifth postulate, that an exterior angle of a triangle has a measure greater than either opposite interior angle, and the statement also holds for hyperbolic geometry. The familiar relationship of measures of exterior and opposite angles is modified only slightly for omega triangles.

THEOREM 9.4. For any omega triangle $AB\Omega$, the measures of the exterior angles formed by extending $\overline{AB}$ are greater than the measures of their opposite interior angles.

This theorem may be proved indirectly by eliminating the other

FIGURE 9.10

FIGURE 9.11

two possibilities. Suppose, in Figure 9.10, that $m(\angle CA\Omega) < m(\angle AB\Omega)$. Then a point D on $\overleftrightarrow{A\Omega}$ can be found such that $\angle CAD \cong \angle ABD$. But this is impossible, since $\triangle ABD$ is an ordinary triangle and the exterior angle cannot be congruent to an opposite interior angle. Suppose next, as in Figure 9.11, that $\angle CA\Omega \cong \angle AB\Omega$. Let D be the midpoint of $\overline{AB}$, let $\overline{DE}$ be perpendicular to $\overrightarrow{B\Omega}$ and let $FA = BE$. Then $\triangle FAD \cong \triangle EBD$, FDE is a straight line, and $\angle DFA$ is a right angle. But the angle of parallelism for the distance EF cannot be a right angle because of Theorem 9.2, which means that this is a contradiction and the assumption of congruence of angles must be rejected.

Congruence of omega triangles is somewhat simpler than that of ordinary triangles, since less information is required. One set of conditions for congruence is given in the following theorem.

THEOREM 9.5. Omega triangles $AB\Omega$ and $A'B'\Omega'$ are congruent if the sides of finite length are congruent and if a pair of corresponding angles at A and A' or B and B' are congruent.

The theorem is proved here by assuming that the remaining pair of angles at an ordinary vertex are not congruent, then arriving—on the basis of this assumption—at a contradiction. In Figure 9.12, assume

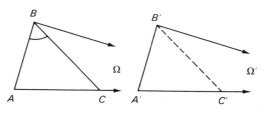

FIGURE 9.12

that $\overline{AB} \cong \overline{A'B'}$ and that $\angle BA\Omega \cong \angle B'A'\Omega'$. Assume that one of the angles, say $\angle AB\Omega$, is greater than $\angle A'v'\Omega'$. Some point C can be located on $\overleftrightarrow{A\Omega}$ such that $\angle ABC \cong \angle A'B'\Omega'$. If C' is located on $\overrightarrow{A'\Omega'}$ so that $\overline{A'C'} \cong \overline{AC}$, then $\triangle ABC \cong \triangle A'B'C'$. But this means that $\angle A'B'C' \cong \angle A'B'\Omega'$, which is a contradiction.

A second set of conditions for the congruence of omega triangles is stated here, but the proof is left as Exercise 7 of Exercise Set 9.2.

THEOREM 9.6. Omega triangles $AB\Omega$ and $A'B'\Omega'$ are congruent if the pair of angles at A and A' are congruent and the pair of angles at B and B' are congruent.

EXERCISES 9.2

1. Explain why a line in hyperbolic geometry must contain two distinct ideal points.

2. Sketch three omega triangles, all with the same ideal vertex and each two of which also have an ordinary vertex in common.

3. Sketch a three-sided figure with two ideal vertices.

4. Sketch a three-sided figure with three ideal vertices.

5. Prove that a line intersecting a side of an omega triangle at a point other than a vertex intersects a second side.

6. Prove that the sum of the measures of the two angles at ordinary vertices of an omega triangle is less than π.

7. Prove Theorem 9.6.

8. Prove that the angle of parallelism is constant for a given distance.

9. Prove that, as the distance increases, the angle of parallelism decreases.

10. Prove that if the two angles at ordinary vertices of an omega triangle are congruent, then the line from the ideal vertex to the midpoint of the opposite side is perpendicular to that side.

11. Prove the converse of the statement in Exercise 10.

9.3   QUADRILATERALS AND TRIANGLES

It should not be surprising to find that omega triangles have some peculiar properties, since they are not identical to any sets of points studied in Euclidean geometry. It is even more significant, however, to consider how the concept of omega triangles, developed in Section 9.2, leads very logically to the formulation of seemingly strange theorems for ordinary triangles and quadrilaterals in hyperbolic geometry.

Among the attempts to prove Euclid's fifth postulate, the most productive were those using the indirect method. By adopting a contradictory postulate and reaching valid conclusions based upon it, mathematicians were actually developing non-Euclidean geometry, even though they remained unaware of the significance of their work.

Girolamo Saccheri (1667–1733), in his attempt to prove the fifth postulate, made use of a set of points now called a *Saccheri quadrilateral.* A Saccheri quadrilateral has two right angles and two congruent sides, as shown in Figure 9.13. $\overline{AB}$ is called the *base* and

FIGURE 9.13

$\overline{CD}$ is called the *summit* of the quadrilateral. The two congruent segments are the *sides.* The next theorem shows that some properties of the corresponding figures in Euclidean geometry continue to hold in hyperbolic geometry.

THEOREM 9.7. The segment joining the midpoint of the base and summit of a Saccheri quadrilateral is perpendicular to both.

In Figure 9.14, $\triangle DAE \cong \triangle CBE$, so $\triangle DFE \cong \triangle CFE$, and the proof follows. Note that Euclid's work on congruence of triangles,

FIGURE 9.14

which does not depend on the fifth postulate, continues to hold in non-Euclidean geometry.

Some of the properties of the Saccheri quadrilateral are unlike any for sets of points in Euclidean geometry.

THEOREM 9.8. The summit angles of a Saccheri quadrilateral are congruent and acute.

The congruence of the summit angles is a consequence of the congruence of the pairs of triangles in Figure 9.14. The significant fact that the summit angles are acute is a consequence of established properties of the omega triangle. In Figure 9.15, $m(\angle EC\Omega) > m(\angle CD\Omega)$,

FIGURE 9.15

since $\angle EC\Omega$ is an exterior angle for omega triangle $CD\Omega$. Since $\angle AD\Omega \cong \angle BC\Omega, m(\angle BCE) > m(\angle ADC)$. But $\angle ADC \cong \angle BCD$, and therefore $m(\angle BCE) > m(\angle BCD)$, so $\angle BCD$ is acute.

J. H. Lambert (1728–1777), like Saccheri, attempted to prove the fifth postulate by an indirect argument. He began with a quadrilateral with three right angles, now called a Lambert quadrilateral, shown in Figure 9.16.

FIGURE 9.16

THEOREM 9.9. The fourth angle of a Lambert quadrilateral is acute.

The proof of Theorem 9.9 follows from the observation that, in Figure 9.14, $EFCB$ and $EFDA$ are Lambert quadrilaterals.

Theorem 9.9 is needed to prove the following even more significant theorem, one that serves to clearly distinguish between hyperbolic and Euclidean geometry.

THEOREM 9.10. The sum of the measures of the angles of a right triangle is less than π.

In Figure 9.17, let $\triangle ABC$ be any right triangle, with D the midpoint of the hypotenuse. $\overline{DE}$ is perpendicular to $\overline{BC}$. Line $\overleftrightarrow{AF}$ is

FIGURE 9.17

constructed so that $\angle FAD \cong \angle EBD$ and $\overline{AF} \cong \overline{BE}$. Then $\triangle AFD \cong \triangle BED$. This means that $\angle AFE$ is a right angle and $\angle ADF \cong \angle EDB$, so E, D, F is a straight line. The consequence is that $ACEF$ is a Lambert quadrilateral with the acute angle at A. The two angles at A, $\angle CAB$ and $\angle BAF$, have the sum of their measures equal to the sum of the measures of $\angle CAB$ and $\angle CBA$, so the sum of the measures of the three angles of $\triangle ABC$ is less than π

The following two theorems are corollaries of Theorem 9.10. The proofs are left as Exercises 8 and 9 of Exercise Set 9.3.

THEOREM 9.11. The sum of the measures of the angles of any triangle is less than π.

DEFINITION. The difference between π and the angle sum is called the *defect*.

For example, if the angle sum were $19\pi/20$, the defect would be $\pi/20$.

THEOREM 9.12. The sum of the measures of the angles of any convex quadrilateral is less than 2π.

You have found, in general, that the theory of congruence of triangles in hyperbolic geometry is much like the theory of congruence in Euclidean geometry. One significant difference results from an application of Theorem 9.12.

THEOREM 9.13. Two triangles are congruent if the three pairs of corresponding angles are congruent.

The proof is by contradiction. Assume, as in Figure 9.18, that $\triangle ABC$ and $\triangle ADE$ have three pairs of corresponding angles congruent. As a result, quadrilateral $BCED$ has the sum of the measures of its angles equal to 2π, but this is a contradiction of Theorem 9.12. This means the assumption of the existence of the similar but noncongruent triangles must be rejected for hyperbolic geometry.

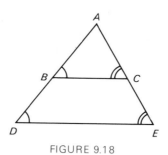

FIGURE 9.18

One consequence of Theorem 9.13 is that, in hyperbolic geometry, the shape and size of triangles are not independent. All triangles of the same shape are necessarily of the same size. Similar but noncongruent figures do not exist in hyperbolic geometry. The existence of a pair of such triangles is equivalent to Euclid's fifth postulate.

EXERCISES   9.3

1. What is the maximum number of angles of a Saccheri quadrilateral that could be congruent to each other?

2. In hyperbolic geometry, why can there be no squares or rectangles?

3. Show that, for a figure such as Figure 9.19, if $AD > BC$, then $m(\angle BCD) > m(\angle ADC)$.

FIGURE 9.19

4. For a Lambert quadrilateral, which is longer, a side adjacent to the acute angle or the side opposite? Prove that your answer is correct.

5. Which is longer, the base or the summit of a Saccheri quadrilateral?

6. Prove that your answer to Exercise 5 is correct.

7. Let a triangular region be partitioned into two triangular regions by a segment through a vertex. Compare the defect of the original triangle with the defect of the two smaller triangles formed.

8. Prove Theorem 9.11.

9. Prove Theorem 9.12.

10. State and prove a theorem giving minimum conditions that must be known if two Saccheri quadrilaterals are to be congruent.

11. State and prove a theorem giving a different set of minimum conditions than your answer to Exercise 10.

9.4   PAIRS OF LINES AND AREA OF TRIANGULAR REGIONS

Two parallel lines in hyperbolic geometry meet in an ideal point because the concept of ideal point was defined in that way. While the proofs are not included in this text, it is possible to prove that if one line is parallel to a second, then the second is parallel to the first, and that if two lines are both parallel to a third in the same sense, they are also parallel to each other.

One more type of point must be created for a complete discussion of the set of points in hyperbolic geometry.

DEFINITION. Two nonintersecting lines are said to meet at a *gamma point* (Γ). Another name for this point is an *ultra-ideal point*.

Recall that nonintersecting lines do not include the parallel lines through a point to a line. For example, if $\overleftrightarrow{AC}$ and $\overleftrightarrow{AB}$ are the parallels through A to $\overleftrightarrow{DE}$, then $\overleftrightarrow{AF}$, $\overleftrightarrow{AG}$, and $\overleftrightarrow{AH}$ are three of the infinite number of nonintersecting lines through A. Each of these lines has a gamma point in common with $\overleftrightarrow{DE}$ (Figure 9.20).

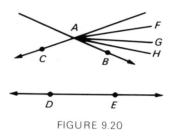

FIGURE 9.20

Besides having a gamma point in common, nonintersecting lines have, perhaps surprisingly, one of the properties of parallel lines in Euclidean geometry: they have a common perpendicular.

THEOREM 9.14. Two nonintersecting lines have a common perpendicular.

In the analysis figure shown in Figure 9.21, it can be seen that

FIGURE 9.21

$\overline{AB}$ and $\overline{CD}$, on two nonintersecting lines, can be though of as the base and summit of Saccheri quadrilateral $ABCD$; hence $\overline{EF}$, connecting the midpoints of the base and summit, is the common perpendicular. The problem is reduced to finding two congruent segments, such as $\overline{AD}$ and

$\overline{BC}$, both perpendicular to one of the given nonintersecting lines, since the common perpendicular can then be found.

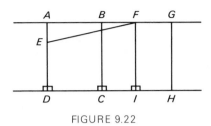

FIGURE 9.22

Assume, as in Figure 9.22, that $\overline{AD}$ and $\overline{BC}$ are perpendicular to the same line, but not congruent. Let $\overline{ED} \cong \overline{BC}$, and $\angle FED \cong \angle FBC$. If $\overline{BG} \cong \overline{EF}$ and $\overline{CH} \cong \overline{DI}$, then quadrilateral $BGHC$ can be shown to be congruent to $EFID$, and $\overline{GH}$ is perpendicular at H; thus, $\overline{BC}$ and $\overline{GH}$ are the required sides for the Saccheri quadrilateral. Note that the proof of the existence of point F on line AB has not been included here.

Because of Theorem 9.14, all of the lines perpendicular to a given line can be said to have the same ultra-ideal point in common. It is left as an exercise to show that two lines in hyperbolic geometry cannot have two distinct common perpendiculars.

With the inclusion of the previous information about the gamma point and the common perpendicular for nonintersecting lines, Table 9.2 can be given summarizing the relationships between pairs of lines intersecting at each kind of point in hyperbolic geometry.

TABLE 9.2

Type of Point Common to Two Lines	Variation in Distance
Ordinary point	Lines diverge from their point of intersection
Ideal point	Lines converge in the direction of parallelism and diverge in the opposite direction
Ultra-ideal point	Lines diverge from their common perpendicular

You may find it interesting to observe that in none of the cases are the two lines always equidistant. You may speculate about the nature of a set of points with each member the same distance from a given line (see Section 9.5). Notice also in this connection that defining parallel lines as two lines everywhere equidistant is equivalent to assuming the fifth postulate.

The fact that there is no square in hyperbolic geometry means that a method of measuring the area of a plane region must be devised that does not depend on square units. Congruent triangles exist in hyperbolic geometry, and the theory of area can be based on this concept in a way very similar to the modern theory of area in Euclidean geometry.

DEFINITION. Two polygons are called *equivalent* if they can be partitioned into the same finite number of pairs of congruent triangles.

For example, polygons *ABCD* and *CDEF* in Figure 9.23 are equivalent but not congruent. Two polygons both equivalent to another

FIGURE 9.23

polygon are also equivalent to each other. Recall that in hyperbolic geometry, the difference between π and the sum of the measures of a triangle is defined as the *defect* of the triangle. The connection between the defect and equivalence of triangles is made clear in the following theorem.

THEOREM 9.15. Two triangles are equivalent if and only if they have the same defect.

If two triangles are equivalent, they can be partitioned into a finite number of pairs of congruent triangles. The defect of each of the original triangles is equal to the sum of the defects of the triangles in the partitioning; hence, the original defects are equal. See Exercise 7, Exercise Set 9.4.

Now suppose that two triangles have the same defect. If they also have a pair of corresponding sides congruent, they can be shown to be equivalent to congruent Saccheri quadrilaterals and hence equivalent to each other.

FIGURE 9.24

For example, in Figure 9.24, let triangles ABC and JHI have the same defect and congruent sides $\overline{BC}$ and $\overline{IJ}$. Triangle ABC is equivalent to Saccheri quadrilateral $BCGF$, where D and E are the midpoints of $\overline{AC}$ and $\overline{AB}$ (see Exercise 8, Exercise Set 9.4). Triangle HIJ is equivalent to Saccheri quadrilateral $IJNM$, where K and L are the midpoints of $\overline{HJ}$ and $\overline{HI}$. But the two Saccheri quadrilaterals are congruent because they have congruent summits and congruent summit angles.

Finally, suppose that two triangles have the same defect, but no pair of congruent sides. Let triangles ABC and DEF be any two such triangles, as in Figure 9.25, with $DF > AC$.

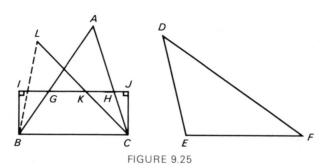

FIGURE 9.25

If G and H are the midpoints of $\overline{AB}$ and $\overline{AC}$, then $\triangle ABC$

is equivalent to Saccheri quadrilateral *BIJC*. Point *K* on $\overline{IJ}$ is a point located so that $KC = \frac{1}{2}DF$, and $\overline{LC} \cong \overline{DF}$. Triangles *ABC* and *LBC* can be shown to have the same defect and to be equivalent. Triangles *LBC* and *DEF* also have the same defect and a pair of congruent sides, hence are equivalent. Since $\triangle DEF$ and $\triangle ABC$ are both equivalent to the same triangle, they are equivalent to each other.

In the proof of Theorem 9.15, the significance of the defect of a triangle was that it made possible the equivalence of triangles to Saccheri quadrilaterals that could be proved congruent. Since the equivalence of triangles depends on the defect, it is possible to define the measure of area of a triangle in hyperbolic geometry as *kd*, where *d* is the defect and *k* is a positive constant the same for all triangles in hyperbolic geometry. The value of *k* depends on the particular triangle chosen to have a unit area. It should be evident that the measure of area increases as the defect increases. In other words, the larger the triangle, the smaller the sum of the measures of the angles.

Oddly enough, in hyperbolic geometry, triangles do not become larger and larger without limit. The triangle with three ideal vertices,

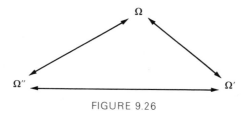

FIGURE 9.26

shown in Figure 9.26, is the *triangle of maximum area*, and its angle sum could be considered zero.

EXERCISES   9.4

1. Explain why two distinct lines cannot have more than one common perpendicular.

2. Where are two nonintersecting lines closest together?

3. How many gamma points are on each line in hyperbolic geometry?

4. Do two lines in hyperbolic geometry always diverge from their point of intersection?

5. Prove that two parallel lines converge continuously in the direction of parallelism.

6. Prove that two polygonal regions equivalent to the same polygonal region are equivalent to each other.

7. Prove that the defect of a triangle is equal to the sum of the defects of the triangles formed by partitioning the original, using the median from one vertex.

8. Prove that $\triangle ABC$, Figure 9.24, is equivalent to Saccheri quadrilateral $BCGF$.

9. Why, in Figure 9.24, do the two Saccheri quadrilaterals have congruent summit angles?

10. Make a sketch to show how a hexagon and a triangle could be equivalent.

11. Show how, starting with any triangle, a triangle of maximum area can be drawn containing the given triangle in its interior.

9.5   CURVES

In hyperbolic geometry, much of the Euclidean theory of circles must be modified because it depends on the fifth postulate. For example, the measure of an inscribed angle is no longer equal to half the measure of its intercepted arc. At the same time, the definition of circles and those properties relating to circles that do not depend on the fifth postulate, such as some having to do with perpendicularity, are parts of hyperbolic geometry.

Defining ideal and ultra-ideal points so that each two lines in hyperbolic geometry have a point in common leads to applications that are generalizations involving circles.

For any two points A and B on an ordinary circle, as in Figure 9.27a, $\angle OAB \cong \angle OBA$, where O is the center of the circle,

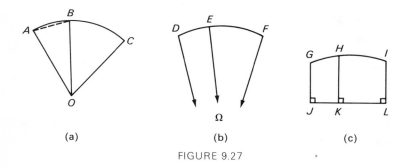

(a)   (b)   (c)

FIGURE 9.27

because $\triangle ABO$ is isosceles. This leads to a generalized definition of *corresponding points*.

DEFINITION. Two points, one on each of two lines, are called corresponding points if the two lines form congruent angles on the same side with the segment whose endpoints are the two given points.

Thus, A and B in Figure 9.27a are corresponding points because of the congruence of angles OAB and OBA. Points C and B, or C and A, similarly can be shown to be corresponding points. In fact, any two points on a circle are corresponding points. An ordinary circle may be defined as the set of all points corresponding to a given point (not the center) on a pencil of rays with an ordinary point as center. A pencil of rays is all rays in a plane with a common endpoint. Because of the fact that the tangent to a circle is perpendicular (orthogonal) to the radius at the point of contact, a circle can be considered the *orthogonal trajectory* of a pencil of rays with an ordinary vertex.

Two new curves, considered generalized circles in hyperbolic geometry, may now be defined.

DEFINITION. A *limiting curve* (see Figure 9.27b) is the set of all points corresponding to a given point on a pencil of rays with an ideal point as vertex.

A limiting curve may be considered the orthogonal trajectory of a pencil of rays with an ideal vertex.

DEFINITION. An *equidistant curve* (see Figure 9.27c) is the set of all points corresponding to a given point on a pencil of rays with a common perpendicular.

An equidistant curve may be considered the orthogonal trajectory of a pencil of rays with a common perpendicular.

A limiting curve in hyperbolic geometry has many of the properties of an ordinary circle. For example, a line perpendicular to a chord at its midpoint is a radius. Theorem 9.16 states a second common property.

THEOREM 9.16. Three distinct points on a limiting curve uniquely determine it.

In Figure 9.28, the perpendicular bisectors of $\overline{AB}$ and $\overline{BC}$, for A, B, C on a limiting curve, determine the unique point Ω, the center of the limiting curve. It should be emphasized that a part of the given condition was that the three points were on some limiting curve. This is not necessarily the case for any three points chosen at random. For example, in Figure 9.28, three points A, B, C lie on an ordinary circle if the perpendicular bisectors at D and E meet at a real point.

FIGURE 9.28

The following theorem states a property of limiting curves that is unlike any property of circles.

THEOREM 9.17. Any two different limiting curves are congruent.

Let $ABCD$ and $A'B'C'D'$ in Figure 9.29 be any two limiting curves. Suppose that points A, B, C, D are given and that A' is any

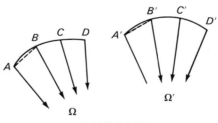

FIGURE 9.29

point on a second limiting curve. If $\angle B'A'\Omega'$ is constructed congruent to $\angle BA\Omega$, and if $\overline{A'B'}$ is congruent to $\overline{AB}$, then point B' is located so that omega triangle $A'B'\Omega'$ is congruent to omega triangle $AB\Omega$. The result is that $\angle AB\Omega \cong \angle A'B'\Omega'$, and B' and A' are corresponding points with B' on the second limiting curve. In the same way, it can be shown that points $C', D', \ldots$ on the limiting curve can be found so that $\overline{B'C'} \cong \overline{BC}$, $\overline{C'D'} \cong \overline{CD}$, and so on. A corollary of Theorem 9.17 is the statement that congruent chords intercept congruent arcs and congruent arcs are intercepted by congruent chords for limiting curves.

The second new type of curve to be studied in hyperbolic geometry is the equidistant curve. Let A, B, C be any three points on an equidistant curve, as shown in Figure 9.30. The common perpendicular,

FIGURE 9.30

$\overleftrightarrow{DF}$, is called the *baseline*. Quadrilateral $ABED$ is a Saccheri quadrilateral, since A and B are corresponding points and $\angle BAD \cong \angle ABE$. The name "equidistant curve" is justified because every point on the curve is the same perpendicular distance from the baseline.

All the points in the plane the same distance from the baseline actually lie on an equidistant curve of two branches, as shown in Figure 9.31. In Exercise Set 9.5, properties of the figure $ABCD$ that are surprisingly like those of a parallelogram are investigated. The equidistant curve has many other properties, a few of which are stated here. Though

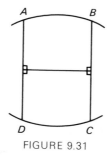

FIGURE 9.31

it is not as fundamental as the limiting curve in the analytic treatment of non-Euclidean geometry, it is simpler to study.

Three points on an equidistant curve determine the curve uniquely, since the baselines can be determined. Not all equidistant curves are congruent—only those whose points are the same distance from the baseline. For congruent equidistant curves, congruent chords intercept congruent arcs, and congruent arcs are intercepted by congruent chords. Given one line, an intersecting line diverges from the point of intersection; in addition, parallel lines and nonintersecting lines are at a nonconstant distance. It is only the equidistant curve in hyperbolic geometry that has the equidistant property characteristic of parallel lines in Euclidean geometry.

The statement that a circle can always be found that passes through three noncollinear points is equivalent to the fifth postulate of Euclid. In hyperbolic geometry, on the other hand, three noncollinear points may lie on a circle, a limiting curve, or on one branch of an equidistant curve.

EXERCISES   9.5

1. In hyperbolic geometry, is the measure of an inscribed angle less than or greater than half the measure of its intercepted arc?

2. Why are any two given points on an ordinary circle corresponding points?

3. Prove that the segments of radii between any pair of limiting curves with the same ideal center are congruent.

4. Could a straight line intersect a limiting curve in three distinct points? Why?

5. Could a straight line intersect one branch of an equidistant curve in three distinct points? Why?

6. Show how to construct the baseline of an equidistant curve, given three points on one branch of the curve.

7. Prove that any segment connecting two points, one on each branch of an equidistant curve, is bisected by the baseline.

8. Explain how to construct the baseline of an equidistant curve given three points, not all of which are on the same branch of the curve.

9. Show that three different equidistant curves pass through the vertices of a triangle, with two vertices on one branch and the third vertex on the other branch.

10.   Compare the properties of *ABCD* in Figure 9.31 with those of a parallelogram in Euclidean geometry.

9.6   ELLIPTIC GEOMETRY

The first five sections of this chapter have included a look at many of the major ideas of hyperbolic non-Euclidean geometry. Not long after the development of hyperbolic geometry, the German mathematician Riemann (1826–1866) suggested a geometry, now called *elliptic,* based on the alternative to the fifth postulate, which states that there are no parallels to a line through a point on the line.

CHARACTERISTIC POSTULATE OF ELLIPTIC GEO-METRY. Any two lines in a plane meet at an ordinary point.

Various finite geometries from Chapter 1, such as the three-point geometry and Fano's geometry, satisfy the requirement of this postulate, although they do not satisfy all the postulates of elliptic geometry.

It is also necessary to further modify the postulational system of Euclid by replacing the statement about infinitude of a line with the milder statement that a line is *boundless* in elliptic geometry. An intuitive idea of the meaning of the word boundless is that the line cannot be enclosed by a circle lying in the same plane. That is, boundless means unbounded, as it was used in Chapter 3.

If Figure 9.32 is in elliptic geometry, with $\overline{EH}$ and $\overline{FI}$ both perpendicular to $\overleftrightarrow{CG}$, these two lines meet at some point A because

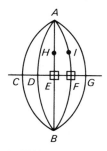

FIGURE 9.32

of the characteristic postulate that any two lines intersect. In elliptic geometry, it is customary to use curves to show straight lines. If $\overline{DE} \cong \overline{EF}$, then $\triangle AEF \cong \triangle AED$, and $\overline{AD}$ is also perpendicular to $\overline{CG}$. By an extension of this argument, it can be shown that every line through A intersects $\overleftrightarrow{CG}$ at right angles. Point A is called a *pole* of $\overleftrightarrow{CG}$ and the line is the *polar* for point A. Here, the distance from A to any point on $\overleftrightarrow{CG}$ is a constant. Recall other uses of the words pole and polar to show a relationship between a point and a line in various geometries encountered in this text.

In Figure 9.32, if $\overline{AE} \cong \overline{EB}$, then $\triangle BEF \cong \triangle AEF$, and A, F, and B are also collinear. This means that B is also a pole of $\overleftrightarrow{CG}$, and that two lines intersect in two points. It is assumed here that A and B are distinct points, although it is also possible to consider them as identical (in *single elliptic geometry*). The common polar of the two points of intersection is the unique common perpendicular to the two lines. Interestingly enough, two straight lines, such as ADB and AFB, enclose a region in elliptic geometry. This region is called a *digon* or *biangle*.

Fortunately, the elliptic geometry of the plane can be explained conveniently by comparison with a familiar model, the earth and lines of longitude on its surface. In Figure 9.32, think of A and B as the north and south poles, and $\overleftrightarrow{CG}$ as the equator. Be very careful to observe that the geometry on the surface of a sphere is *not* non-Euclidean but instead provides a three-dimensional model for two-dimensional elliptic geometry.

In Figure 9.33, let $\overleftrightarrow{HI}$ and $\overleftrightarrow{KJ}$ be lines in elliptic geometry

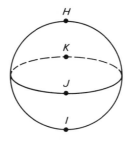

FIGURE 9.33

such that H and I are the poles of $\overleftrightarrow{KJ}$, and K and J are the poles of $\overleftrightarrow{HI}$. In elliptic geometry, the distance from any line to its pole is constant—the same for all lines. Furthermore, a line is of finite length, and the length is four times the distance from the pole to the line.

The use of a model also helps to explain what it means when a line is called boundless. A great circle on the sphere, representing a line in elliptic geometry, cannot be enclosed by a curve on the sphere. There is no way to "get around" the great circle from a point on one side of it to a point on the other without intersecting the great circle.

In elliptic geometry, there are no parallel or nonintersecting lines, since any two lines meet. However, there are quadrilaterals and triangles that have some properties analogous to those encountered in hyperbolic geometry.

THEOREM 9.18. The segment joining the midpoint of the base and summit of a Saccheri quadrilateral is perpendicular to both the base and the summit.

The proof of this theorem is identical to the proof of Theorem 9.7 for hyperbolic geometry.

THEOREM 9.19. The summit angles of a Saccheri quadrilateral are congruent and obtuse.

In Figure 9.34, let $ABCD$ be any Saccheri quadrilateral, with O and O' the poles of $\overleftrightarrow{EF}$, the line joining the midpoints of the base and summit.

FIGURE 9.34

The fact that the summit angles are congruent comes from the congruent triangles used in the proof of Theorem 9.18.

To show that the summit angles are obtuse, it can first be established that their complements are acute. If, as in Figure 9.34, X lies on BO' and is the pole of $\overleftrightarrow{BC}$, then $BX > BO'$, since $BO' < EO'$. This means that $\angle BCX$ is a right angle, so $\angle BCO'$ is acute and its supplement, $\angle BCD$, is obtuse.

The following theorems are consequences of Theorem 9.19, and the proofs are left as Exercises 10–12 of Exercise Set 9.6.

THEOREM 9.20. In elliptic geometry, a Lambert quadrilateral has its fourth angle obtuse, and each side of this angle is shorter than the side opposite.

THEOREM 9.21. The sum of the measures of the angles of any triangle is greater than π.

THEOREM 9.22. The sum of the measures of the angles of any quadrilateral is greater than 2π.

You should find it worthwhile to return to the sections on hyperbolic geometry to see which of the concepts not already mentioned can be used in elliptic geometry. For example, there is no angle of parallelism in elliptic geometry. Circles do exist in elliptic geometry. They can be described as the set of all points a fixed distance from a given point. Since each point on a circle is also the same distance from the polar for the given point, a circle can be considered as an equidistant curve in elliptic geometry.

EXERCISES   9.6

1. Name two other geometries from Chapter 1 for which the characteristic postulate of elliptic geometry holds.

2. In ordinary Euclidean geometry, is a line boundless?

3. A line of latitude that is not the equator of a sphere represents what other concept in the model of elliptic geometry shown in Figure 9.33?

4. In elliptic geometry, if the distance from a line to its pole has a measure of two, what is the measure of the length of a line in that geometry?

5. When may two points not always determine a unique line in elliptic geometry?

6. Do limiting curves exist in elliptic geometry? Why?

7. What is the maximum measure of the third angle in a triangle in elliptic geometry, if two of the angles are right angles?

8. Verify the fact that the proof of Theorem 9.18 is the same as that for Theorem 9.7.

9. In elliptic geometry, where are two given lines farthest apart?

10. Prove Theorem 9.20.

11. Prove Theorem 9.21.

12. Prove Theorem 9.22.

13. Prove that similar but noncongruent triangles cannot exist in elliptic geometry.

9.7 CONSISTENCY OF NON-EUCLIDEAN GEOMETRY

The question of which kind of geometry is the "right" geometry to fit the physical universe is one that may never be answered. While it might seem a simple problem to solve, there are formidable difficulties. For example, it might seem that physical measurement of the angles of triangles would easily settle the question of whether the sum is or is not equal to π. But physical measurement always involves errors. Besides, it is known that in non-Euclidean geometry the angle sum depends on the size of the triangle. The small part of the universe in which we live may not be big enough to contain triangles with a defect (or excess) large enough to be measured. Space travel may someday settle the question.

For most practical purposes, it makes little difference in our lives whether the universe is Euclidean or not. Euclidean geometry provides a simple model to use in most practical applications, such as engineering. Einstein's general theory of relativity asserts that physical space that is in the neighborhood of any kind of matter is best described by the postulates of elliptic geometry. Non-Euclidean geometry may have new applications as man explores more of the universe. In many cases, it could be used instead of Euclidean geometry without making much difference.

From a purely mathematical point of view, the truth of an axiomatic system is not what must be investigated. The important thing to determine is whether or not the system is *consistent*, as discussed briefly in Chapter 1. In other words, do the axioms of non-Euclidean geometry lead to *valid* conclusions, without any contradictions. Actually, the mathematician is interested in what is called *relative consistency*. He needs to be sure that non-Euclidean geometry is as consistent as Euclidean geometry or as the algebra of real numbers.

Beltrami is given the credit for first proving the consistency of non-Euclidean geometry, in 1868. The proof of the relative consistency of a non-Euclidean geometry consists of finding a model within Euclidean geometry that, with suitable interpretations, has the same postulational structure as the non-Euclidean geometry. Then any inconsistency in the non-Euclidean geometry would mean there is also an inconsistency in Euclidean geometry.

It has already been mentioned that the lines of longitude on a sphere provide a three-dimensional model for lines in plane elliptic geometry. More generally, great circles in the model correspond to lines, and each line has two poles associated with it. Lines of latitude are models for equidistant curves.

Two-dimensional models in Euclidean geometry for hyperbolic geometry include those suggested by Felix Klein and Poincaré. The Klein model is illustrated in Figure 9.35. Points on the circle represent ideal points. Secants of the circle represent hyperbolic lines, with points on the chords interior to the circle representing ordinary points. In Figure 9.35, $\overleftrightarrow{AB}$ and $\overleftrightarrow{CD}$ are intersecting lines, $\overrightarrow{AD}$ and $\overleftrightarrow{CD}$ are parallel lines, and $\overleftrightarrow{AD}$ and $\overleftrightarrow{BC}$ are nonintersecting lines.

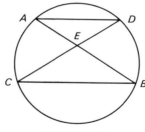

FIGURE 9.35

One advantage of the Klein model is that straight lines are represented by portions of straight lines. Two major problems in connection with the Klein model soon become apparent, however. One is that some interpretation of distance must be given so that a segment such as $\overline{AD}$ in Figure 9.35 has the characteristics of a line of infinite length. The second major problem is that some interpretation of angle must be given to avoid having the sum of the measures of the angles of a triangle equal to π.

The Poincaré model eliminates the second problem named in the previous paragraph, but not the first. The relationship between the

Poincaré and the Klein models for hyperbolic geometry is illustrated in Figure 9.36. On a plane tangent to the sphere, Klein's model is shown as the interior of the circle C' congruent to the great circle C'' of the sphere.

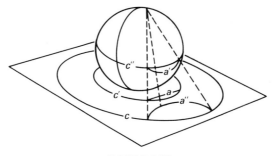

FIGURE 9.36

The points on C' and its interior are projected onto the bottom hemisphere of the sphere by a projection from an ideal point. Each chord, such as a, is projected into the arc a' of a circle orthogonal to C''. Now, the points on the bottom hemisphere are carried by a stereographic projection (considered in Section 6.4 as an application of inversive geometry) into the points on circle C and its interior. The arc a' is projected into an arc a'' orthogonal to circle C in the tangent plane. Poincaré's model for ordinary points in hyperbolic geometry is illustrated by the interior of C. The ordinary points on a line are illustrated by the points on an arc, such as a'', interior to C.

According to the development above, lines in Poincaré's model are represented by arcs of circles orthogonal to a given circle. In Figure 9.37, $\overset{\frown}{AB}$ and $\overset{\frown}{FG}$ represent intersecting lines, $\overset{\frown}{AB}$ and $\overset{\frown}{BC}$ represent parallel lines, and $\overset{\frown}{AB}$ and $\overset{\frown}{DE}$ represent nonintersecting lines.

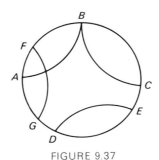

FIGURE 9.37

The ingenious way in which distance is defined in Poincaré's model is shown in the following formula, which refers to Figure 9.38.

$$CD = k \log_e \frac{\dfrac{AC}{CB}}{\dfrac{AD}{DB}},$$

where AC, CB, AD, and DB are lengths of segments, not arcs, and k is a parameter. Check to see that this formula does yield the correct results when the distance between two points is infinite (such as for D and B in Figure 9.38) or is zero—that is, if the two points coincide. Also note that the concept of cross ratio used in the formula is familiar from the related concept studied in the projective geometry of Chapter 7.

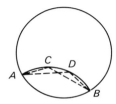

FIGURE 9.38

Fortunately, the measure of angles in Poincaré's model may be defined in the usual way for Euclidean geometry. This is true because angles are preserved under a stereographic projection. Inversion is used as a key tool to show that the geometry of points and lines in Poincaré's model has the same postulational structure as the geometry of the hyperbolic plane. For example, a proof by inversion is used to prove that, in the Poincaré model, the sum of the measures of the angles of a triangle is less than π.

In Figure 9.39a, triangle ABC, which represents any triangle in the Poincaré model, is inverted into $A'B'C'$, shown in Figure 9.39b, by an inversion with center O, the inverse point of point B, with the fixed circle as the circle of inversion. It is apparent from the shape of $A'B'C'$, since arc $A'C'$ bends inward, that the sum of the angle measures is less than π. It is important to note that Figure 9.39 does not actually show the correct relative positions of the set of points in (a) and its inverse (b), since this relative position is unimportant in the proof. Actually the sets of points would overlap if drawn correctly, but such a drawing would be confusing, thus making it harder to interpret relationships within the inverted figure.

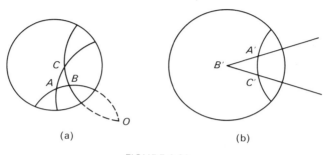

(a)   (b)

FIGURE 9.39

The relatively brief discussion of consistency of non-Euclidean geometries in this section has used several concepts from earlier chapters. Each of the many geometries presented in this book has importance in its own right, but each also contributes to a better understanding of the other geometries and to a fuller appreciation of the meaning of the expression "modern geometries." Other modern geometries, especially those more dependent on an analytic approach, remain to be explored by the interested reader.

EXERCISES 9.7

Exercises 1–4 refer to the Klein model.

1. Draw a triangle with three ordinary points as vertices.
2. Draw an omega triangle.
3. Draw a triangle with exactly two ideal vertices.
4. Draw a triangle with three ideal vertices.

Exercises 5–8 refer to the Poincaré model.

5. Draw an example of an angle of parallelism.

6. Draw an omega triangle.

7. Draw a triangle with exactly two ideal vertices.

8. Draw a triangle with three ideal vertices.

9. In Figure 9.38, find the distance AD, using the formula given.

10. In the proof by inversion about the measures of angles, explain in detail how Figure 9.39b was derived from Figure 9.39a.

HILBERT'S AXIOMS

The axioms in Appendix 1 are reprinted from **David Hilbert**, *The Foundations of Geometry,* La Salle: Open Court Publishing Company, 1950, by permission of Open Court Publishing Company. A new tenth edition of *The Foundations of Geometry* was published by Open Court Publishing Company in 1971.

GROUP I. AXIOMS OF CONNECTION

I, 1. Two distinct points A and B always completely determine a straight line a. We write $AB = a$ or $BA = a$.

I, 2. Any two distinct points of a straight line completely determine that line; that is, if $AB = a$ and $AC = a$, where $B \neq C$, then is also $BC = a$.

I, 3. Three points A, B, C not situated in the same straight line always completely determine a plane α. We write $ABC = \alpha$.

I, 4. Any three points A, B, C of a plane α, which do not lie in the same straight line, completely determine that plane.

I, 5. If two points A, B of a straight line a lie in a plane α, then every point of a lies in α.

I, 6. If two planes α, β have a point A in common, then they have at least a second point B in common.

I, 7. Upon every straight line there exist at least two points, in every plane at least three points not lying in the same straight line, and in space there exist at least four points not lying in a plane.

GROUP II.   AXIOMS OF ORDER

II, 1. If A, B, C are points of a straight line and B lies between A and C, then B lies also between C and A.

II, 2. If A and C are two points of a straight line, then there exists at least one point B lying between A and C and at least one point D so situated that C lies between A and D.

II, 3. Of any three points situated on a straight line, there is always one and only one which lies between the other two.

II, 4. Any four points A, B, C, D of a straight line can always be so arranged that B shall lie between A and C and also between A and D, and furthermore, that C shall lie between A and D and also between B and D.

II, 5. Let A, B, C be three points not lying in the same straight line and let a be a straight line lying in the plane ABC and not passing through any of the points A, B, C. Then, if the straight line a passes through a point of the segment AB, it will also pass through either a point of the segment BC or a point of the segment AC.

GROUP III.   AXIOM OF PARALLELS

III.   In a plane α there can be drawn through any point A, lying outside of a straight line a, one and only one straight line which does not intersect the line a. This straight line is called the parallel to a through the given point A.

GROUP IV.   AXIOMS OF CONGRUENCE

IV, 1. If A, B are two points on a straight line a, and if A' is a point upon the same or another straight line a', then, upon a given side of A' on the straight line a', we can always find one and only one point B' so that the segment AB (or BA) is congruent to the segment $A'B'$. We indicate this relation by writing $AB \equiv A'B'$. Every segment is congruent to itself; that is, we always have $AB \equiv AB$.

IV, 2. If a segment AB is congruent to the segment $A'B'$ and also to the segment $A''B''$, then the segment $A'B'$ is congruent to the segment $A''B''$; that is, if $AB \equiv A'B'$ and $AB \equiv A''B''$, then $A'B' \equiv A''B''$.

IV, 3. Let AB and BC be two segments of a straight line a which have no points in common aside from the point B, and, furthermore, let $A'B'$ and $B'C'$ be two segments of the same or of another straight line a' having, likewise, no point other than B' in common. Then, if $AB \equiv A'B'$, and $BC \equiv B'C'$, we have $AC \equiv A'C'$.

IV, 4. Let an angle (h, k) be given in the plane α and let a straight line a' be given in a plane α'. Suppose also that, in the plane α', a definite side of the straight line a' be assigned. Denote by h' a half-ray of the straight line a' emanating from a point O' of this line. Then in the plane α' there is one and only one half-ray k' such that the angle (h, k) or (k, h) is congruent to the angle (h', k') and at the same time all interior points of the angle (h', k') lie upon the given side of a'. We express this relation by means of the notation $\angle (h, k) \equiv \angle (h', k')$. Every angle is congruent to itself; that is

$$\angle (h, k) \equiv \angle (h, k),$$

$$\angle (h, k) \equiv \angle (k, h).$$

IV, 5. If the angle (h, k) is congruent to the angle (h', k') and to the angle (h'', k''), then the angle (h', k') is congruent to the angle (h'', k''); that is to say, if $\angle (h, k) \equiv \angle (h', k')$ and $\angle (h, k) \equiv \angle (h'', k'')$ then $\angle (h', k') \equiv \angle (h'', k'')$.

IV, 6. If, in the two triangles ABC and $A'B'C'$, the congruences $AB \equiv A'B'$, $AC \equiv A'C'$, $\angle BAC \equiv \angle B'A'C'$ hold, then the congruences $\angle ABC \equiv \angle A'B'C'$ and $\angle ACB \equiv \angle A'C'B'$ also hold.

GROUP V.   AXIOM OF CONTINUITY

V.   Let A_1 be any point upon a straight line between the arbitrarily chosen points A and B. Take the points A_2, A_3, A_4,... so that A_1 lies between A and A_2, A_2 between A_1 and A_3, A_3 between A_2 and A_4, etc. Moreover, let the segments AA_1, $A_1 A_2$, $A_2 A_3$, $A_3 A_4$,... be equal to one another. Then, among this series of points, there always exists a certain point A_n such that B lies between A and A_n.

BIRKHOFF'S POSTULATES

These axioms are reprinted from G. D. Birkhoff, "A Set of Postulates for Plane Geometry (based on scale and protractor)," *Annals of Mathematics,* Vol. 33, 1932, by permission of the *Annals of Mathematics.*

I. *Postulate of line measure:* The points $A, B, \ldots$ of any line l can be put into $(1, 1)$ correspondence with the real numbers x so that $|x_B - x_A| = d(A, B)$ for all points A, B.

II. *Point-line postulate:* One and only one straight line l contains two given points P, Q $(P \neq Q)$.

III. *Postulate of angle measure:* The half-lines $l, m, \ldots$ through any point O can be put into $(1, 1)$ correspondence with the real numbers a

(mod 2π) so that, if $A \neq O$ and $B \neq O$ are points on l and m, respectively, the difference $a_m - a_l$ (mod 2π) is $\angle AOB$. Furthermore, if the point B on m varies continuously in a line r not containing the vertex O, the number a_m varies continuously also.

IV. *Postulate of similarity:* If in two triangles, $\triangle ABC$, $\triangle A'B'C'$, and for some constant $k > 0$, $d(A', B') = kd(A, B)$, $d(A', C') = kd(A, C)$ and also $\angle B'A'C' = \pm\angle BAC$, then also $d(B', C') = kd(B, C)$, $\angle C'B'A' = \pm\angle CBA$, and $\angle A'C'B' = \pm\angle ACB$.

POSTULATES FROM HIGH SCHOOL GEOMETRY

The postulates in Appendix 3 are reprinted from H. R. Pearson and J. R. Smart, *Geometry,* Boston: Ginn and Company, 1971, by permission of Ginn and Company.

1. Space exists and contains at least two distinct points.
2. If two points are distinct, then there is exactly one line which contains them.
3. Every line is a set of points and contains at least two distinct points.
4. No line contains all of the points of space.
5. If a point is in a line and another point is not in that line, then the two points are distinct.

6. If three points are distinct and noncollinear, then there is exactly one plane which contains them.

7. Every plane is a set of points and contains at least three distinct noncollinear points.

8. No plane contains all of the points of space.

9. If two distinct planes intersect, then their intersection is a line.

10. If two distinct points of a line lie in a plane, then every point of the line lies in the plane.

11. There exists a correspondence which associates the number one with an arbitrarily chosen pair of distinct points and a unique positive real number with every other pair of distinct points.

12. There is a one-to-one mapping of the real numbers onto the points in a line such that 0 and 1 are mapped onto the points O and U, respectively. The measure of the distance between any two points in the line is the absolute value of the difference of their corresponding numbers.

13. If P and Q are distinct points in line L and p and q are distinct real numbers, then there is a unique coordinate system for L which assigns P the coordinate p, and Q the coordinate q.

14. Any line in a plane separates the points of the plane which are not points of the line into two sets such that
    1. Each of the two sets is a convex set, and
    2. Every segment which joins a point of one set to a point of the other intersects the line.

15. Any plane separates the points of space which are not in the plane into two sets such that
    1. Each of the two sets is a convex set, and
    2. Every segment which joins a point of one set to a point of the other intersects the plane.

16. There exists a correspondence which associates with each angle in space exactly one real number n such that $0 < n < 180$.

17. For every point O and every closed half-plane whose edge contains O, there is a one-to-one mapping of the real numbers n, where $0 \leq n \leq 180$, onto the set of all rays in the closed half-plane having O as their endpoint.

18. For any angle ABC there exists exactly one ray-coordinate system with $\overrightarrow{BA}$ as the zero-ray such that for any point X in the C-side of $\overleftrightarrow{BA}$, $\overrightarrow{BX}$ corresponds to a real number $n, 0 < n < 180$.

19. If I is the interior of $\angle ABC$, R is the set of interior points of all rays between $\overrightarrow{BA}$ and $\overrightarrow{BC}$, H is the set $AB/C \cap BC/A$, and S is the set of interior points of segments which join a point in $\overset{\circ}{\overrightarrow{BA}}$ and a point in $\overset{\circ}{\overrightarrow{BC}}$, then $I = R = H = S$.

20. If for a one-to-one mapping of the vertices of one triangle onto the vertices of another (not necessarily distinct from the first), two sides and the included angle of one are congruent to the corresponding two sides and the included angle of the other triangle, then the two triangles are congruent.

21. If for a one-to-one mapping of the vertices of one triangle onto the vertices of another (not necessarily distinct from the first), two angles and the included side of one triangle are congruent to the corresponding two angles and the included side of the other, then the two triangles are congruent.

22. If for a one-to-one mapping of the vertices of one triangle onto the vertices of another (not necessarily distinct from the first), the three sides of one triangle are congruent to the corresponding three sides of the other, then the two triangles are congruent.

23. If a point is not in a given line, then there is no more than one line containing the point and parallel to the given line.

24. There exists a correspondence which associates the number one with an arbitrarily chosen polygonal region and a unique positive real number with every polygonal region.

25. If the polygonal region R is the union of two polygonal regions R_1 and R_2, whose interiors do not intersect, then relative to a given unit area, the measure of the area of R is the sum of the measures of the areas of R_1 and R_2.

26. The measure of the area of a square region is the square of the measure of the length of its side.

27. If two triangles are congruent, then the triangular regions bounded by the triangles have the same area.

28. There exists a correspondence which associates the number one with an arbitrarily chosen geometric solid and a unique positive real number with every other geometric solid.

29. If the polyhedral solid S is the union of two polyhedral solids S_1 and S_2 whose interiors do not intersect, then the measure of the volume of S is the sum of the measures of the volumes of S_1 and S_2.

30.   If a polyhedral solid whose boundary is a rectangular parallel-opiped has B, h, and V as the measures of the area of its base, its altitude, and its volume, respectively, then $V = Bh$.

31.   Given two geometric solids S_1 and S_2 and a plane R, if every plane parallel to R which intersects either S_1 or S_2 also intersects the other, and if the intersections are regions with equal area, then the volumes of S_1 and S_2 are equal.

32.   If two polyhedral solids are congruent, they have equal volumes.

EXAMPLES OF NOTATION USED IN TEXT

Addition of coordinates of points	$X + A$
Angle	$\angle AOA'$
Closed neighborhood	$N[P, r]$
Congruence	$\overline{IY} \cong \overline{IX}$
Distance between two points	AB
Element and image under mapping	$x \rightarrow x + 2$
Element of	$\in$
Harmonic set of points	$H(AB, CD)$
Ideal point in hyperbolic geometry	Ω
Intersection	$\cap$
Inverse transformation	f^{-1}
Line	$\overleftrightarrow{AB}$

Measure of angle $m \angle ABC$

Open neighborhood $N(P, r)$

Permutation symbol $\begin{pmatrix} 1 & 2 & 3 \\ 3 & 1 & 2 \end{pmatrix}$

Perspectivity $ABCD \overset{s}{\overline{\wedge}} A'B'C'D'$

Projectivity $ABCD \overline{\wedge} A'B'C'D'$

Proper or improper subset $K' \subseteq K$

Ray $\overrightarrow{AD}$

Reflection R_1

Rotation $R(O, \alpha)$

Segment $\overline{PP'}$

Segment without endpoints $\overset{\circ}{\overline{PP}}{}^{\circ}$

Segment without one endpoint $\overset{\circ}{\overline{PP'}}$ or $\overline{PP}{}^{\circ}$

Set-builder notation $\{(x, y) \mid x > 3, x, y \in R\}$

Similar to $\triangle ABC \sim \triangle DEF$

Subset $\subset$

Ultra-ideal point Γ

Vector $\boldsymbol{AA'}$

ANSWERS TO SELECTED EXERCISES

EXERCISES 1.1

1. a. $3^2 + 4^2 = 5^2$; b. 30   **2.** $\approx 7{,}787$ miles   **3.** False   **4.** True
5. False   **6.** True   **7.** True   **8.** False

EXERCISES 1.2

8. II, 5   **9.** I, 1 and I, 2   **10.** II, 1–4   **11.** 11   **12.** 12   **13.** 13
14. 14 and 15   **15.** 20, 21, 22   **16.** 24, 25, 26, 27   **17.** Yes   **18.** Yes
19. Yes   **20.** Yes

EXERCISES 1.3

2. (1) There exist exactly three distinct books in the system.   (2) Two distinct

books are in exactly one library. (3) Not all the books in the system are in the same library. (4) Two distinct libraries have at least one book in common.
3. (1) There exist exactly three different students. (2) Two different students are on exactly one committee. (3) Not all the students are on the same committee. (4) Two different committees have at least one student in common.
4. None   **5.** Two   **6.** No   **7.** No   **8.** No

EXERCISES 1.4

1. (1) There exist exactly three distinct lines in the geometry. (2) Two distinct lines are on exactly one point. (3) Not all the lines of the geometry are on the same point. (4) Two distinct points are on at least one line.
3. None
4. (1) The total number of committees is four. (2) Each pair of committees has exactly one student in common. (3) Each student is on exactly two committees.
5. No   **6.** Four   **7.** None   **9.** Axiom 2
10. (1) The total number of trees is four. (2) Each pair of trees has exactly one row in common. (3) Each row contains exactly two trees.
11. No   **12.** One

EXERCISES 1.5

1. (1) There exists at least one library. (2) Every library has exactly three books in it. (3) Not all books are in the same library. (4) For two different books, there is exactly one library containing both of them. (5) Each two libraries have at least one book in common.
2. 1, 3, 4   **3.** 4, 5, 6;   4, 5, 3;   4, 5, 2;   4, 5, 7;   4, 1, 2;   4, 1, 3;   4, 1, 6;
4, 1, 7;   4, 2, 3;   4, 2, 7;   4, 6, 3;   4, 6, 7   **4.** 4, 5, 1 and 7, 6, 1;   4, 6, 2 and
5, 2, 7;   4, 7, 3 and 5, 6, 3
6. (1) There exists at least one row. (2) Every row has exactly three trees in it. (3) Not all trees are in the same row. (4) There exists exactly one row containing a tree not on a row that contains no tree of the given row. (5) If a tree is not on a row, there exists exactly one different tree in the row such that the two trees do not have a row in common. (6) With the exception in Axiom 5, exactly one row contains each pair of distinct trees.
7. 1, 3, 4   **8.** Two

EXERCISES 1.6

2. a. A;   b. B' **3.** a. $\overleftrightarrow{A'P}$;   b. $\overleftrightarrow{PB'}$ **4.** BRB' and ASA' **5.** 1, 3, 4, 5
6. $\overleftrightarrow{A'T}, \overleftrightarrow{B'P}, \overleftrightarrow{B'R}$ **10.** Four   **11.** $a, b,$ and d **12.** a. 21;   b. 31;   d. 57
15. Four   **16.** Ten, five   **17.** None   **18.** Two   **19.** Four, four
20. One   **21.** Twelve, six   **22.** One

EXERCISES 2.1

2. c **3.** a. $(7, -2)$ b. $(5, -3)$ c. $(2, -5)$

4. $\{(a,i),(c,j),(e,k)\}$ **5.** Not defined
6. $\{(b,a),(d,c),(h,e)\}$ **7.** $\{(i,b),(j,d),(k,h)\}$
8. $(x,y) \rightarrow (x-3, y-1)$ **9.** $(x,y) \rightarrow (x-3, y-1)$
10. $(x,y) \rightarrow (x+5, y-2)$ **11.** $(x,y) \rightarrow (x-2, y+3)$

EXERCISES 2.2

3. R_1 **4.** R_2 **5.** $R(240)$ **6.** $R(120)$ **7.** No
8.

	R_1	I
R_1	I	R_1
I	R_1	I

10. R_1, I; I **13.** One subgroup with eight members; three subgroups with four members; five subgroups with two members; and one subgroup with one member.

EXERCISES 2.4

1. a. $(3, -5)$; b. $(8, 2)$; c. $(2, -13)$; d. $(1, -13/2)$
2. a. $(-5, -9)$; b. $(0, 0)$ **3.** $(-\sqrt{2}/2, 7\sqrt{2}/2)$
4. $(\sqrt{2}/2, 3\sqrt{2}/2)$ **5.** $x - h = (x' - h)\cos\alpha + (y' - k)\sin\alpha$,
$y - k = -(x' - h)\sin\alpha + (y' - k)\cos\alpha$ **6.** $\left(\dfrac{3 + 7\sqrt{3}}{2}, \dfrac{3\sqrt{3} - 7}{2}\right)$
7. $\left(\dfrac{17\sqrt{3} - 3}{2}, \dfrac{3\sqrt{3} + 3}{2}\right)$ **8.** $(-5, 10)$ **9.** $x = x' + 5$; $y = y' - 2$
10. $x = x'\cos 30° + y'\sin 30°$; $y = -x'\sin 30° + y'\cos 30°$
11. $x = x'$; $y = -y'$ **12.** $x = x'\cos 60° + y'\sin 60°$;
$y = x'\sin 60° - y'\cos 60°$

EXERCISES 2.5

6. The result is that a and b in the general equations for the translation are zero, so the translation is the identity.   **7.** Only the center of rotation is an invariant point.   **9.** Let the lines be $y = 0$ and $y = mx$.

EXERCISES 2.6

1. a. $(-2, 1, 7)$; b. $(0, 6, 10)$; c. $(-4, -2, 2)$; d. $(-2, -3, 8)$
2. a. $(1, 3, -4)$; b. $(3, 2, -1)$;   **4.** $(-3\sqrt{2}/2, 4, 7\sqrt{2}/2)$
5. $(3, 8, -1)$ **6.** $(-2, -4, 3)$ **7.** $(-2, 3, 9)$ **8.** $(7, 4, -2)$
11. a. Reflection, rotatory reflection, or glide reflection; b. Reflection, rotatory reflection, or glide reflection   **12.** $x' = x, y' = y, z' = 4 - z$

EXERCISES 2.7

1. 4 inches   **2.** 6/5   **3.** Length and area   **4.** The measure of area is multiplied by the square of the ratio of similarity.   **5.** $\sqrt{51/6}$
6. They are equal.   **7.** (9/4, 15/4)   **8.** (28, 30)   **10.** Yes

EXERCISES 3.1

1. b, c, d, e, f, g   **2.** a, b   **4.** Yes; no   **8.** The open one-dimensional neighborhood with radius r of a point P is the set of interior points of a segment of length $2r$ with P as midpoint.   **9.** e. The empty set, the set itself, the complement of the set.   f. The empty set, the entire plane, the empty set.
10. c. The empty set, $\{(x, y, 0): x \geq y\}$, the complement of the boundary.   d. The empty set, the set itself, the complement of the set.   **11.** a. Closed; open; open; closed; closed; neither   b. Closed; closed; neither; closed   c. Neither
12. a. a, d;   b. a, b   **15.** It is neither closed nor open.   **16.** a. The interior of a circular region.   b. A parabola and its interior.

EXERCISES 3.2

1. No   **2.** Yes   **3.** Yes   **4.** No   **5.** All regular   **6.** Vertices are corner points.   **7.** Vertex is corner point.   **8.** Vertices are corner points.   **9.** a. A triangular region is the intersection of the three supporting half-planes determined by the sides of the triangle.   b. A convex polygonal region is the intersection of the n supporting half-planes determined by the sides of the polygon.
11. If a line contains at least one boundary point of the set but no interior points, it is a supporting line for a convex set of points. If a line is not a supporting line for a convex set of points (but does contain at least one boundary point), then it contains interior points. If a line contains at least one boundary point of the set and also contains interior points, then the line is not a supporting line for a convex set of points.

EXERCISES 3.3

1. Yes   **2.** No, not bounded   **3.** No, not bounded   **4.** No, not convex
5. No, not bounded   **6.** No, not always bounded   **7.** No, not bounded
8. Yes   **12.** Line segment   **14.** Three of the first and six of the second
15. Two hundred Black Angus and one hundred Hereford

EXERCISES 3.4

1. Door handle and door, book on a shelf   **2.** It fills an octant in space, with the vertex at the origin and three sides lying on the reference planes.   **3.** No. It must pass through a boundary point of the convex set.   **4.** Yes   **5.** No
6. No   **7.** Yes   **8.** Yes   **10.** A tetrahedral solid is the intersection of the four closed half-spaces containing the sides of the tetrahedron and the vertex not on the side.

EXERCISES 3.5

1. a. Circular region;   b. Triangular region;   c. Entire plane;   e. Parabolic region; f. Entire plane;   g. Angle and interior;   h. Triangular region or quadrangular region   **2.** Packing a carton of books; getting a large, irregularly shaped package through a door   **4.** a. Yes;   b. Yes   **6.** One branch of a hyperbola and the point of intersection of the two asymptotes   **8.** a. The angle and its interior;   c. A spherical region;   d. A tetrahedral region or other special cases **9.** No

EXERCISES 3.6

1. 1 inch, $\sqrt{2}$ inches   **2.** 2 inches, $\sqrt{13}$ inches   **3.** 1 inch, 1 inch
4. $3\sqrt{187/14}, 7$ **9.** $4\pi, 8(\pi - \sqrt{3})$ **10.** $\sqrt{3}$

EXERCISES 3.7

6. Yes   **7.** Let $S = \{P_1, \ldots, P_n\}$ be any finite set of points in a plane. Then there is a point A such that every closed half-plane formed by a plane through A contains at least $n/3$ points of S.

EXERCISES 4.1

1. ABC, H; ACH, B; AHB, C; CHB, A **2.** At the vertex of the right angle
3. For an obtuse triangle   **4.** Three   **5.** Yes

EXERCISES 4.2

1. $\dfrac{AC}{CB} \cdot \dfrac{BD}{DF} \cdot \dfrac{FE}{EA} = -1$ **2.** $\dfrac{CA}{AB} \cdot \dfrac{BF}{FD} \cdot \dfrac{DE}{EC} = -1$ **7.** $\dfrac{AE}{EC} \cdot \dfrac{CF}{FB} \cdot \dfrac{BD}{DA} = 1$
12. An altitude

EXERCISES 4.3

1. a. One;   b. One;   c. An infinite number

EXERCISES 4.4

1. A median   **3.** At the centroid   **4.** No   **5.** Shorter   **6.** 3/4; 3/5; 4/5

EXERCISES 4.5

1. Yes   **8.** The four points are vertices of a square.   **9.** The four points are vertices of a square.

EXERCISES 4.6

1. 1.6180   **2.** $55/34 \approx 1.618$ **3.** $1/x^4$ or $5 - 3x$ **6.** The interior angles have measures of 135, and this number is not a factor of 360.

EXERCISES 5.2

12. $\dfrac{(c - a)(fg - eh) - (g - e)(bc - ad)}{(c - a)(h - f) - (g - e)(d - b)}$,

$$\frac{(d - b)}{(c - a)}\left[\frac{(c - a)(fg - eh) - (g - e)(bc - ad)}{(c - a)(h - f) - (g - e)(d - b)}\right] + \frac{bc - ad}{c - a}$$

EXERCISES 5.3

3. There are 4, 3, 2, 1, or no solutions.   **4.** 1 solution   **5.** 1 or no solution
6. 1 or no solution   **7.** 1 or no solution   **8.** 2, 1, or no solutions
9. 4 solutions   **10.** 4, 3, 2, 1, or no solutions   **11.** 2, 1, or no solutions
12. 1 solution   **13.** 2, 1, or no solutions   **14.** 1 solution   **15.** 1 or no
solution   **16.** 1 or no solution   **17.** 1 solution   **18.** 1 or no solution
19. 1 or no solution

EXERCISES 5.4

1. x is a solution of $x^4 - 4x^2 + 2 = 0$.   **2.** Yes, it is a solution of $x^3 - 2 = 0$.
3. ≈ 2.5198 **5.** The only possibilities are ± 1, $\pm 1/2$, $\pm 1/4$, $\pm 1/8$, and each of
these can be checked by synthetic division to show it is not a solution.   **8.** None
9. This makes the cosine of an acute angle equal to 1.   **10.** The equation has
no rational solution.

EXERCISES 6.1

1. 16/3   **2.** 3   **3.** 4/3   **4.** 1/3   **5.** 8/49   **6.** $\sqrt{5}/3$ **7.** 189/50
8. No   **9.** Outside the circle of inversion   **10.** When the circle does not pass
through the center of inversion; when the circle does pass through the center
of inversion.   **12.** The identity   **13.** Two intersecting circles, two intersecting
lines, an intersecting line and circle   **14.** A circle inside and concentric to the
circle of inversion   **15.** On the circle of inversion

EXERCISES 6.2

1. Two orthogonal circles, a circle and a line through the center   **10.** A line not
passing through the point

EXERCISES 6.3

1. (3/25, 4/25)   **2.** (2, 2)   **3.** (0, 3)   **4.** (56/25, 8/25)   **5.** (16/17, -4, 17)

6. (27/13, 18/13)   **8.** $2x^2 + 2y^2 - 9x = 0$ **9.** $5x^2 + 5y^2 - 9x = 0$
10. $2x^2 + 2y^2 + 9x + 9y = 0$ **11.** $36x^3 + 36xy^2 - 81y^2 = 0$
12. $9y^3 + 9x^2y - 81x^2 = 0$ **13.** $x^2 + y^2 = 81/16$ **15.** $\dfrac{2 - 5i}{29}$

16. $\dfrac{3 + 7i}{58}$ **17.** $\dfrac{4 - 8i}{5}$ **18.** $\dfrac{27 + 99i}{130}$

EXERCISES 6.4

1. The other point would be the ideal point.   **3.** A tangent to a circle passes through the center of a circle orthogonal to the circle at the point of tangency.
4. A circle circumscribed about a right triangle has the hypotenuse as a diameter.
13. The same plane

EXERCISES 7.1

1. $A'B' = AB\cos\theta$ **2.** a. Top center; b. Off to the left   **4.** Not invariant
5. Invariant   **6.** Not invariant   **7.** Not invariant   **8.** Not invariant
9. Not invariant   **10.** Invariant   **11.** Not invariant

EXERCISES 7.2

1. There are fewer invariant properties.   **2.** 1, 2, 3, 5, 7   **3.** 1, 3   **4.** 1, 3
5. 1, 2   **6.** 1, 2, 3, 4, 7   **7.** 1, 2   **8.** 1, 2   **9.** 1, 2   **11.** Two   **12.** Two
13. Two of them would be ideal points.   **14.** Yes   **15.** One

EXERCISES 7.3

1. If A and B are distinct points of a plane, there is at least one line on both points.   **2.** On a line are at least four points.   **3.** A triangle (trilateral) consists of three nonconcurrent lines and the points of intersection of each pair.
4. Three points not on the same line determine a plane.   **5.** A point is determined by two intersecting lines.

EXERCISES 7.4

3. $H(XC, DY)$; $H(CX, YD)$; $H(XC, YD)$ **4.** A point at infinity   **5.** Yes
6. Yes   **7.** Yes   **8.** Yes   **9.** Yes

EXERCISES 7.6

1. The last coordinate is zero.   **2.** a. (3, 8, 1); b. (2, 3/4, 1); c. (1, −4, 1)
3. a. (2, 5) b. (−1/2, 3/2); c. (2/3, −5/3)   **4.** (5/3, 2/3, 1); (10, 4, 6); (20, 8, 12)
5. No   **6.** (1, −1, 0)   **7.** $3x_1 + x_2 - 2x_3 = 0$ **8.** $3X_1 + X_2 - 2X_3 = 0$
9. $x_1 - x_3 = 0$ **10.** (1, 0, −1)   **11.** $x_1 - x_3 = 0$ **12.** (0, 2, 1)

EXERCISES 7.7

1. $-2, 3, 1$ **2.** $(5, -4)$ **3.** $(-4, -4)$ **4.** $(1, 1, -1)$; $(1, 1, 1)$; $(1, 1, 0)$; $(4, 4, -1)$ **5.** $5/3$ **7.** Similarity   **8.** None   **9.** Affine   **10.** Motion

EXERCISES 7.8

1. The product of a reflection and the same reflection is the identity. **2.** a. Rotation of $180°$; b. Rotation of $120°$; c. Rotation of $90°$ **5.** A unique one-dimensional involution is determined by two pairs of corresponding points. **6.** A two-dimensional projectivity that leaves the four vertices of a complete quadrangle invariant is the identity transformation.

EXERCISES 7.9

1. It could be a degenerate conic.   **2.** Figures 7.5, 7.19, 7.21, 7.22, 7.23, 7.25 **3.** The lines of the ranges of points in the projectivity defining the line conic are also lines of the conic.   **4.** The points corresponding to the common point of the two ranges of points determining a line conic are the points of contact of the two ranges of points.   **5.** $3x_1 + x_2 + x_3 = 0$ **7.** Yes   **9.** There are 720 permutations of six things taken six at a time. Half of these are duplicates caused by reversing the order of points. Of the remaining 360, there are six different names for each of 60 hexagons.

EXERCISES 8.1

1. A rotation; $x \rightarrow$ cotangent x **2.** The inverse also has an inverse that is a continuous transformation.   **4.** Length of a segment; area of a region; number of sides of a polygon   **5.** A convex set is a connected set for which the curve joining two points can always be a segment.   **6.** a, b, c, d, f, h   **7.** a, c, d, e **8.** a. Penny; b. Washer; c. Button   **10.** a

EXERCISES 8.2

1. a, d   **2.** a, b, c   **3.** Vertex C is not counted.   Vertex E is counted. **4.** Odd   **5.** Simply connected   **8.** Yes

EXERCISES 8.3

1. No. Images would not always be in the circular region.   **3.** No   **4.** No **5.** Yes .   **6.** No   **7.** Yes   **8.** 5, 7   **9.** 5   **10.** None

EXERCISES 8.4

2. Zero   **3.** Zero: prism; One: teacup; Two: dining room chair with arms **5.** Four; four   **8.** The theorems only apply to countries topologically equivalent to a circular region.   **9.** More than four countries with a common point would each require a separate color.

EXERCISES 8.5

1. By rows: 6, 12; 20, 30; 12, 20, 30   **5.** 0   **6.** 2; 0; -2; -4 **9.** Two interlocking pieces   **10.** No. Two interlocking pieces   **11.** No. Two interlocking pieces   **12.** Three   **13.** Four

EXERCISES 9.1

3. If a straight line intersects one of two parallel lines, it will not always intersect the other.   **4.** Straight lines parallel to the same straight line are not always parallel to each other.   **5.** There does not exist one triangle for which the sum of the measures of the angles is π radians.   **6.** There does not exist a pair of similar but noncongruent triangles.   **7.** There does not exist a pair of straight lines the same distance apart at every point.   **8.** It is not always possible to pass a circle through three noncollinear points.   **9.** a
10. AGJ; AIL; AKM **12.** Two

EXERCISES 9.2

1. Otherwise, two distinct lines could be both parallel and intersecting.

EXERCISES 9.3

1. Two   **2.** The sum of the measures of the angles would be 2π.   **4.** Adjacent side   **5.** Summit   **7.** The defect of the original triangle is equal to the sum of the defects of the two smaller triangles.

EXERCISES 9.4

1. There would exist a rectangle.   **2.** At their common perpendicular   **3.** An infinite number   **4.** No   **9.** The sum of the measures of the summit angles is equal to the sum of the measures of the three angles of the original triangles.

EXERCISES 9.5

1. Less than   **2.** The angles determined by the radii and the segment joining the points are congruent.   **4.** No   **5.** No   **8.** Connect the midpoints of two of the segments joining pairs of points.

EXERCISES 9.6

1. Four-line geometry; $PG(2,3)$ **2.** Yes   **3.** Circle   **4.** 8   **5.** When they

are the two poles of a line.   **6.** No, there are no ideal points.   **7.** Less than 2π
9. At the polar of their common points

EXERCISES 9.7

9. Infinite   **10.** $\overleftrightarrow{A'B'}$ is the image of circle A, B. $\overleftrightarrow{B'C'}$ is the image of circle B, C. Circle A', C' is the image of circle A, C. Each of the three images is orthogonal to the image of the fundamental circle.

BIBLIOGRAPHY

Adler, Claire F., *Modern Geometry,* Second Edition. New York: McGraw-Hill, 1967.

Barry, Edward H., *Introduction to Geometrical Transformations.* Boston: Prindle, Weber & Schmidt, 1966.

Benson, Russell V., *Euclidean Geometry and Convexity.* New York: McGraw-Hill, 1966.

Blumenthal, Leonard M., *A Modern View of Geometry.* San Francisco: W. H. Freeman, 1961.

Bullard, Sir Edward, The Origin of the Oceans. *Scientific American,* September 1969, Vol. 221 (3).

Chinitz, Wallace, Rotary Engines. *Scientific American,* February 1969, Vol. 220(2).

Chrestenson, H. E., *Mappings of the Plane.* San Francisco: W. H. Freeman, 1966.

Courant, Richard, and Robbins, Herbert, *What Is Mathematics?* London: Oxford University Press, 1951.

Coxeter, H. S. M., *Introduction to Geometry*. New York: Wiley, 1961.

Coxeter, H. S. M., *Projective Geometry*. New York: Blaisdell, 1964.

Davis, David R., *Modern College Geometry*. Reading, Mass.: Addison-Wesley, 1954.

Dorwart, Harold L., *The Geometry of Incidence*. Englewood Cliffs, N.J.: Prentice-Hall, 1966.

Eggleston, H. G., *Convexity*. Cambridge: Cambridge University Press, 1966.

Escher, M. C., *The Graphic Work of M. C. Escher*. New York: Ballantine, 1971.

Eves, Howard, *A Survey of Geometry*, Vol.1. Boston: Allyn and Bacon, 1963.

Fishback, W. T., *Projective and Euclidean Geometry*. New York: Wiley, 1962.

Gans, David, *Transformations and Geometries*. New York: Appleton-Century-Croft, 1969.

Graustein, William C., *Introduction to Higher Geometry*. New York: Macmillan, 1930.

Hilbert, David, *The Foundations of Geometry*. La Salle, Ill., Open Court, 1950.

Johnson, Donovan A., *Paper Folding for the Mathematics Class*. National Council of Teachers of Mathematics, 1957.

Johnson, Donovan A., and Glenn, W. H., *Topology, the Rubber Sheet Geometry*. New York: McGraw-Hill, 1960.

Klee, Victor L. (Ed.), Convexity. Proceedings of the Seventh Symposium in Pure Mathematics of the American Mathematical Society, Providence, R. I., 1963.

Levi, Howard, *Topics in Geometry*. Boston: Prindle, Weber & Schmidt, 1968.

Levy, Lawrence S., *Geometry: Modern Mathematics via the Euclidean Plane*. Boston: Prindle, Weber & Schmidt, 1970.

Lyusternik, L., *Convex Figures and Polyhedra*. New York: Dover, 1963.

Meserve, Bruce E., *Fundamental Concepts of Geometry*. Reading, Mass.: Addison-Wesley, 1955.

Meserve, Bruce E., and Izzo, Joseph A., *Fundamentals of Geometry*. Reading, Mass.: Addison-Wesley, 1969.

Moise, Edwin E., *Elementary Geometry from an Advanced Standpoint*. Reading, Mass.: Addison-Wesley, 1963.

Patterson, E. M., *Topology*. New York: Interscience, 1959.

Pearson, H. R., and Smart, J. R., *Geometry*. Boston: Ginn, 1971.

Pedersen, Jean J., Dressing Up Mathematics. *The Mathematics Teacher,* February 1968, Vol. 61(2).

Perfect, Hazel, *Topics in Geometry*. London: Pergamon Press, 1963.

Pervin, W. J., *Foundations of General Topology*. New York: Academic Press, 1964.

Pratt, M. M., *Finite Geometries*. Unpublished Master's Thesis, San Jose State College, San Jose, California, 1964.

Prenowitz, Walter, and Swain, Henry, *Congruence and Motion in Geometry*. Boston: Heath, 1966.

Rademacher, H., and Toeplitz, O., *The Enjoyment of Mathematics*. Princeton: Princeton University Press, 1957.

Tuller, Annita, *A Modern Introduction to Geometries*. Princeton: Van Nostrand, 1967.

Valentine, F., *Convex Sets*. New York: McGraw-Hill, 1964.

Wolfe, Harold E., *Introduction to Non-Euclidean Geometry*. New York: Dryden Press, 1945.

Wylie, C. R., *Foundations of Geometry*. New York: McGraw-Hill, 1964.

Yaglom, I. M., and Boltyanskiĭ, V. G., *Convex Figures*. New York: Holt, Rinehart & Winston, 1961.

INDEX